COMPUTATIONAL
ECOLOGY
Graphs, Networks and
Agent-based Modeling

COMPUTATIONAL ECOLOGY

Graphs, Networks and Agent-based Modeling

WenJun Zhang

Sun Yat-Sen University, China & International Academy of
Ecology and Environmental Sciences

We **World Scientific**

NEW JERSEY · LONDON · SINGAPORE · BEIJING · SHANGHAI · HONG KONG · TAIPEI · CHENNAI

Published by

World Scientific Publishing Co. Pte. Ltd.

5 Toh Tuck Link, Singapore 596224

USA office: 27 Warren Street, Suite 401-402, Hackensack, NJ 07601

UK office: 57 Shelton Street, Covent Garden, London WC2H 9HE

British Library Cataloguing-in-Publication Data
A catalogue record for this book is available from the British Library.

COMPUTATIONAL ECOLOGY
Graphs, Networks and Agent-based Modeling

ISBN-13 978-981-4343-61-9
ISBN-10 981-4343-61-7

Typeset by Stallion Press
Email: enquiries@stallionpress.com

Printed in Singapore.

In memory of my father, GuoXiang Zhang, and mother, GuiFang Niu

Preface

Networks are mathematically directed (in practical applications also undirected) graphs and a graph is an one-dimensional abstract complex, i.e., a topological space. Network theory focuses on various topological structures and properties, dynamic properties, and functionality-topology relationship of networks, etc. There are some common mathematical foundations, theories and methodology for network analysis, in which graph theory, statistics, and operational research, etc., are the fundamental sciences of network analysis. Various ecological networks, at both micro- and macro- levels, will provide numerous sources for the development of general network theory and methodology and also facilitate the development of theory and methodology of ecological networks.

Ecological network analysis is a fast developing science. Many core scientific issues, for example, ecological structure, co-evolution, co-extinction and biodiversity conservation, etc., are expected to be addressed by network approaches and network analysis. Network analysis is becoming the core methodology to treat complex ecological systems.

In the view of system dynamics, ecological networks are always self-organized systems with emergent, autonomous and adaptive properties. Their dynamics can be represented by agent-based modeling and some other methodologies. Therefore, agent-based modeling falls into the scope of this book. This book is the first comprehensive treatment of the subject in the areas of ecology and environmental sciences. From this integrated and self-contained book, scientists, university teachers and students will be provided with an in-depth and complete insight on knowledge, methodology and recent advances of graphs, networks and agent-based modeling in ecology and environmental sciences. Java codes and a software package, BioNetAnaly, are presented in the book for easy use for those who are not familiar with its mathematical details. Users can find the software package, BioNetAnaly, at http://www.iaees.org/publications/software/index.asp.

I am grateful to the scientists, Professors ShiZe Wang, HonSheng Shang, ZhenQi Li, DeXiang Gu and KG Schoenly for their constant guidance and support. I am also indebted to the people who contributed valuable suggestions, Professors Yi Pang, GuangHua Liu, and ZhiGuo Zhang, *et al.* I would especially like to thank the anonymous reviewers for their comments and suggestions on this book. I thank Mrs YanHong Qi for her help in manuscript preparation, and postgraduates Wu Wei, WenGang Zhou, HeFeng Luo, Hao Zheng, ChunHua Liu, and Na Li for their help in field investigation and manuscript preparation. This book is supported in part by International Academy of Ecology and Environmental Sciences (IAEES).

WenJun Zhang

Contents

PART I

Graphs

Definitions and Concepts

Graph theory is an ancient but also young science (Chan *et al.*, 1982; Gross and Yellen, 2005). Back in the early 18th century, graphs have been used to solve some complex problems. In 1736, Professor Leonhard Euler, a Swiss mathematician, solved the well-known Königsberg Bridges Problem, and published the first paper on graph theory. Until the late and mid-20th century, graph theory was widely known and appreciated, and was treated as an important branch of mathematics.

For discrete structures, graph is a powerful tool. In logistics planning, network and other research, we will encounter various graphs made of vertices and edges.

1. Definitions and Concepts

A graph is a space made of some vertices, and edges that connect the vertices. For example, Fig. 1 shows a graph with 9 vertices and 14 edges. In other words, a graph X is an ordered triple $(V(X), E(X), \varphi)$, or ordered pair $(V(X), E(X))$, in which $V(X)$ is the nonempty vertex set whose elements are vertices, $E(X)$ is the set of edges, whose elements are edges, and φ is the incidence function that associates each edge (e) of X and un-ordered vertices (u, v), and $\varphi(e) = (u, v)$.

Two vertices associated with the same edge are called the adjacent vertices. Two adjacent edges associated with the same vertex are called the adjacency edges. For example, the vertices v_1 and v_2 in Fig. 1 are adjacent; the edges e_6 and e_{10} are adjacent, and therefore they are adjacency edges.

The edge with the same terminal vertex is defined as a loop, that is, given that $\varphi(e) = (u, u)$, the edge is called a loop. The edge with different

3

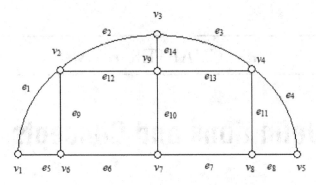

Figure 1.   A graph with 9 vertices and 14 edges.

initial and terminal vertices is defined as a link. There is no loop in Fig. 1 and all edges in Fig. 1 are links.

The vertex without linking any edge is defined as an isolated vertex.

A graph, e.g., Fig. 1, can be expressed as:

$$X = (V(X), E(X), \varphi),$$

$$V(X) = \{v_1, v_2, v_3, \ldots, v_9\},$$

$$E(X) = \{e_1, e_2, e_3, \ldots, e_{14}\},$$

$$\varphi(e_1) = v_1 v_2, \quad \varphi(e_2) = v_2 v_3, \quad \varphi(e_3) = v_3 v_4, \quad \varphi(e_4) = v_4 v_5,$$

$$\varphi(e_5) = v_1 v_6, \quad \varphi(e_6) = v_6 v_7, \quad \varphi(e_7) = v_7 v_8, \quad \varphi(e_8) = v_5 v_8,$$

$$\varphi(e_9) = v_2 v_6, \quad \varphi(e_{10}) = v_7 v_9, \quad \varphi(e_{11}) = v_4 v_8, \quad \varphi(e_{12}) = v_2 v_9,$$

$$\varphi(e_{13}) = v_4 v_9, \quad \varphi(e_{14}) = v_3 v_9$$

or

$$X = (V, E),$$

$$V = \{v_1, v_2, v_3, \ldots, v_9\},$$

$$E = \{e_1, e_2, e_3, \ldots, e_{14}\}$$

$$= \{(v_1, v_2), (v_2, v_3), (v_3, v_4), \ldots, (v_4, v_9), (v_3, v_9)\}.$$

The graph without any vertex and edge is a null graph, denoted by ϕ.

1.1.   *Finite graph and infinite graph*

A graph is called a finite graph given that it has a finite number of vertices (i.e., the vertex set V is a finite set), and a finite number of edges (edge set E is a finite set). Figure 1 is a finite graph.

1.2.   *Simple graph and planar graph*

Given $\varphi(e_1) = \varphi(e_2)$, e_1 and e_2 are called parallel edges. If a graph does not have both loops and parallel edges, it is defined as the simple graph. Figure 1 is a simple graph, while Fig. 2(a) is not a simple graph. There is a loop (e_2) and two parallel edges (e_5, e_6) in Fig. 2(a).

In a graph, X, delete all loops, so that every pair of adjacent edges have only one link, the resulting simple graph is the basic simple graph of graph X.

In a graph, when any two edges do not intersect each other but intersect at the endpoints, then the graph is called a planar graph, or else it is a nonplanar graph. A planar graph can be drawn on a plane in a simple way. Figure 1 is a planar graph, while Fig. 2(b) is a nonplanar graph.

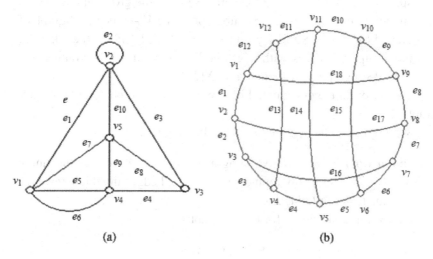

Figure 2.   (a) Non-simple graph. (b) Nonplanar graph.

1.3. *Subgraph, proper subgraph, spanning subgraph, complementary graph*

Given $V(Y) \subseteq V(X)$, $E(Y) \subseteq E(X)$, and φ_Y is the restriction of φ_X on $E(Y)$, Y is called the subgraph of X, denoted by $Y \subseteq X$. In this case, the vertex set and the edge set of Y are subsets of the vertex set and the edge set of X respectively.

For example, in Fig. 1,

$$Y = (V, E),$$
$$V = \{v_1, v_2, v_6\},$$
$$E = \{e_1, e_9, e_5\},$$

is a subgraph.

Given $Y \subseteq X$, and $Y \neq X$, i.e., if the graph Y does not contain all edges of X, then Y is the proper subgraph of X. The graph Y above is the proper subgraph of Fig. 1.

Given $V(Y) = V(X)$, i.e., if the subgraph Y contains all vertices of X, then Y is the spanning subgraph of X. For example, remove the edges e_7 and e_8, the rest of the graph is a spanning subgraph of Fig. 1.

Suppose $V' \subset V$, let V' be the vertex set, and the group of edges with two endpoints in V', the resulting subgraph is the V'-induced subgraph of X, denoted by $X[V']$. Similarly, suppose $E' \subset E$, let E' be the edge set, and the group of vertices with edges in E', the resulting subgraph is the E'-induced subgraph of X, denoted by $X[E']$.

The set of all edges included in X but not in Y is the complementary graph of Y.

Subgraphs can be operated by the following rules:

(1) **Union operation.** For two subgraphs X_1 and X_2 of X, the union $X_1 \cup X_2$, means a subgraph consisting of all edges in which any edge belongs to X_1 or (and) X_2.

For example, in Fig. 1, for the subgraphs

$$X_1 = (V, E),$$
$$V = \{v_1, v_2, v_6\},$$
$$E = \{e_1, e_9, e_5\},$$

and

$$X_2 = (V, E),$$
$$V = \{v_2, v_6, v_7, v_9\},$$
$$E = \{e_9, e_6, e_{10}, e_{12}\},$$

the union is

$$X_1 \cup X_2 = (V, E),$$
$$V = \{v_1, v_2, v_6, v_7, v_9\},$$
$$E = \{e_1, e_5, e_6, e_9, e_{10}, e_{12}\}.$$

(2) **Intersection operation.** For two subgraphs X_1 and X_2 of X, the intersection $X_1 \cap X_2$, means a subgraph consisting of all edges shared by X_1 and X_2. For example, the intersection of above subgraphs X_1 and X_2 is:

$$X_1 \cap X_2 = (V, E),$$
$$V = \{v_2, v_6\},$$
$$E = \{e_9\}.$$

The intersection of two subgraphs is a null graph if they have not shared edges even though they share some vertices.

(3) **Complementary operation.** For two subgraphs X_1 and X_2 of X, the complement $X_1 - X_2$, means a subgraph consisting of all edges belonging to X_1 but not X_2. For example, the complement of the above subgraphs X_1 and X_2 is:

$$X_1 - X_2 = (V, E),$$
$$V = \{v_1, v_2, v_6\},$$
$$E = \{e_1, e_5\}.$$

(4) **Ring sum operation.** For two subgraphs X_1 and X_2 of X, the ring sum $X_1 \oplus X_2$, means a subgraph consisting of all edges specific to either X_1 or X_2, and

$$X_1 \oplus X_2 = (X_1 \cup X_2) - (X_1 \cap X_2)$$

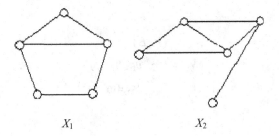

Figure 3.   Two subgraphs.

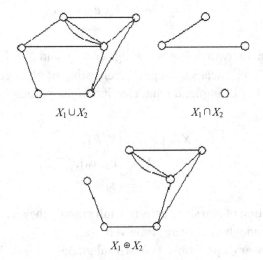

Figure 4.   The union, intersection and ring sum of two graphs X_1 and X_2 in Fig 3.

For example, the ring sum of the above subgraphs X_1 and X_2 is:

$$X_1 \oplus X_2 = (V, E),$$
$$V = \{v_1, v_2, v_6, v_7, v_9\},$$
$$E = \{e_1, e_5, e_6, e_{10}, e_{12}\}.$$

For two subgraphs X_1 and X_2 in Fig. 3, the union, intersection and ring sum are indicated in Fig. 4.

1.4. *Complete graph and m-order complete graph*

A graph is called a complete graph if there exists an edge between any pair of distinct vertices. Figure 1 is not a complete graph but it contains complete subgraphs, such as Y. Every induced subgraph of complete graph is called a complete subgraph.

In a graph, if there are always m edges between any pair of distinct vertices, then the graph is called a m-order complete graph, or m-order regular graph. Figure 1 is not a m-order complete graph, but it contains 1-order complete subgraphs, such as Y.

1.5. *Edge sequence, edge train and path*

If k edges of graph X are naturally arranged then it generates a finited sequence:

$$e_1(v_1, v_2), e_2(v_2, v_3), e_3(v_3, v_4), \ldots, e_k(v_k, v_{k+1}),$$

or

$$v_1 e_1 v_2 e_2 v_3 \cdots v_{k-1} e_k v_k, \quad k \geq 2.$$

This sequence is called a chain, or edge sequence, where k is the chain length. For example, an edge sequence of Fig. 1 is shown as (Fig. 5):

$$e_5(v_1, v_6), e_9(v_6, v_2), e_{12}(v_2, v_9), e_{14}(v_9, v_3), e_3(v_3, v_4), e_4(v_4, v_5),$$
$$e_8(v_5, v_8), e_7(v_8, v_7).$$

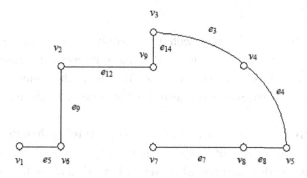

Figure 5. A chain in Fig. 1.

The chain with distinct initial vertex and terminal vertex is called an open chain, or else it is a closed chain. For example, the above chain is an open chain.

A chain without repeated edges is called a simple chain, or edge train. A chain can be an open simple chain or closed simple chain. The chain above is an open simple chain.

An open simple chain without repeated vertices is defined as the elementary chain, or path. The chain above is an elementary chain.

If there is at least a chain starting from initial vertex u to terminal vertex v, then the elementary chain starting from initial vertex u to terminal vertex v exists.

Suppose the two endpoints of an elementary chain are the same vertex, the chain is called a circuit. A circuit with length k is called the k-circuit. According to the parity of k, a k-circuit can be an odd circuit or an even circuit. Graph X contains a circuit if the number of edges is not less than the number of vertices.

1.6.   *Connected graph, unconnected graph, and connected components*

Two vertices, u and v in a graph X is said to be connected, if there exists a path between u and v.

For any given graph X, the vertex set:

$$V = \bigcup_{i=1}^{m} V_i.$$

Two vertices, u and v in V_i is connected, if and only if $u, v \in V_i$. The induced subgraphs $X[V_i]$, $i = 1, 2, \ldots, m$, are defined as the connected components of X. If $m = 1$, X is a connected graph. That is, in the connected graph X, each pair of vertices are connected. Otherwise X is an unconnected graph.

Obviously, X is a connected graph if and only if for each of the classification that divides vertex set V into two nonempty subsets V_1 and V_2, there always exists an edge, and one of its vertex is in V_1 and another vertex is in V_2. Figure 1 is a connected graph.

Given X is a connected graph, then the maximal connected subgraph of X is itself. If X is an unconnected graph, then each of its connected components is the maximal connected subgraph of X.

Assume that the connected graph X has v vertices, its rank is defined as $v - 1$. Suppose the graph X has v vertices and m connected components, then its rank is $v - m$.

Suppose the connected graph X has v vertices and e edges, its nullity is defined as $e - v + 1$. And if X has m connected components, the nullity of X is thus defined as $e - v + m$.

In a connected graph, there exists shared vertex between any two longest elementary chains.

Given the connected graph X has $\omega(X)$ connected components, and the subgraph after deleting an edge e from X is $X - e$, then we have:

$$\omega(X) \leq \omega(X - e) \leq \omega(X) + 1.$$

1.7.   *Separable graph, inseparable graph, bipartite graph and disjoint graph*

If there is a vertex in the connected graph X, and after removing the vertex if the remaining graph becomes an unconnected graph, then the vertex is called the separation vertex.

If there is an edge in the connected graph X, and after removing the edge if the remaining graph becomes an unconnected graph, then the edge is called the bridge.

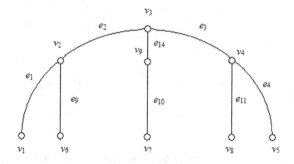

Figure 6.   A tree in Fig. 1.

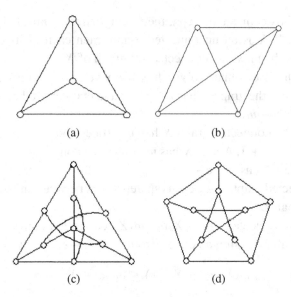

Figure 7.   Two pairs of isomorphic graphs (a) and (b), and (c) and (d).

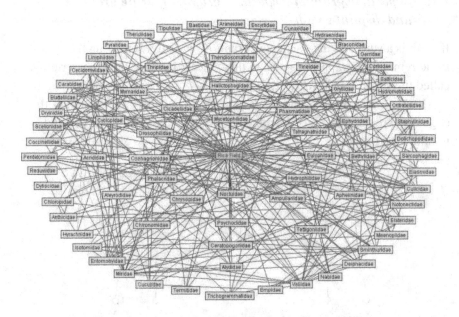

Figure 8.   An ecological interaction graph (Zhang, 2007).

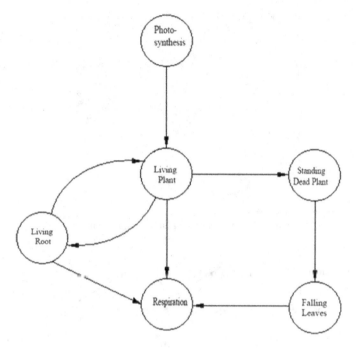

Figure 9.   An ecosystem graph (Zhang, 2007b).

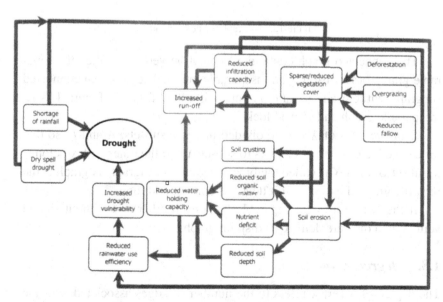

Figure 10.   Drought factors graph (Slegers and Stroosnijder, 2008).

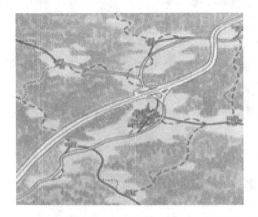

Figure 11.   An ecological landscape network.

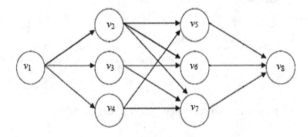

Figure 12.   A graph for pest spread.

The connected graph containing separation vertex is called the separable graph, otherwise it is an inseparable graph. The maximal connected subgraph without any separation vertex is called a block. Figure 1 is an inseparable graph and also a block.

If the vertices of X can be divided into two subsets, A and B, so that for each edge one vertex belongs to subset A and the other vertex belongs to subset B, then X is called the bipartite graph, or bigraph. A graph is the bipartite graph if and only if it does not contain odd loop.

If the two graphs are separable from each other and without shared vertex, then they are defined as disjoint graphs.

1.8. *Degree of vertex*

The degree of a vertex refers to the number of edges associated with the vertex. A loop is equivalent to two edges. A vertex is an isolated vertex if

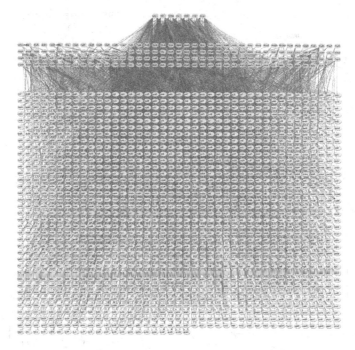

Figure 13. *E. coli* transcriptional regulatory network (Martinez-Antonio, 2011).

its degree is zero. The degree of vertex v_1 in Fig. 1 is two, and the degree of vertex v_9 is four. There is no isolated vertex in Fig. 1.

Suppose the graph $X = (V, E)$, $V = \{v_1, v_2, \ldots, v_n\}$, $E = \{e_1, e_2, \ldots, e_m\}$, the sum of degree of vertex is thus $2m$.

1.9. *Directed graph and undirected graph*

Given that all edges of the graph X are directed, X is called as a directed graph, or else it is an undirected graph. For a directed edge from vertex v_i to v_j, v_i is called to be adjacent to v_j. The edges in Fig. 1 are undirected, so it is an undirected graph.

For the directed graph, the above definitions should be adjusted correspondingly. For the directed graph, e.g., a vertex has outdegree ($d^+(v)$, the number of edges leaving the vertex) and indegree ($d^-(v)$, the number of edges pointing to the vertex). The degree of the vertex, $d(v) = d^+(v) + d^-(v)$.

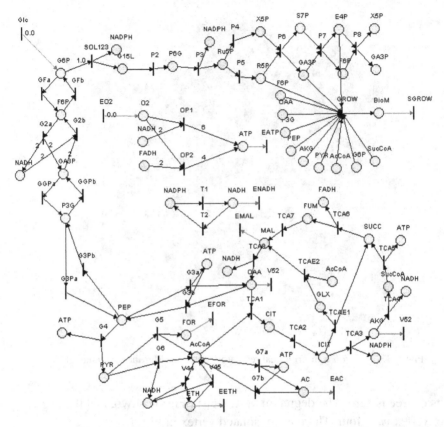

Figure 14.   Metabolic regulatory network in yeast (Luo, 2007).

1.10. *Cutset and association set*

By removing a set of edges from the connected graph X, such that X is separated into two disjoint subgraphs, then the minimal edge set of this kind is defined as the cutset of X.

The set of edges associated with the same vertex is defined as the association set.

1.11. *Tree and tree branch*

For a connected subgraph T of the connected graph X with v vertices, if T has $v - 1$ edges, contains all vertices of X, and does not contain any

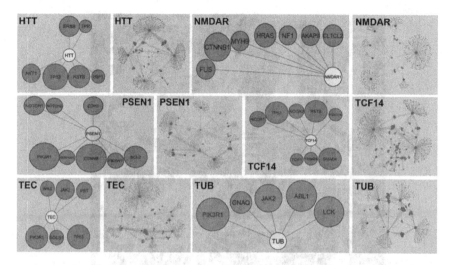

Figure 15. Cancer gene networks (Ibrahim *et al.*, 2011).

circuit, then T is called as a tree of X (Fig. 6). The edges in a tree are called tree branches. A connected graph X may have multiple distinct trees.

2. Isomorphism

If there are two monogamy maps, $\psi : V(X) \to V(Y)$, and $\varphi : E(X) \to E(Y)$, so that $\varphi_X(e) = uv$, if and only if $\varphi_Y(\varphi(e)) = \psi(u)\psi(v)$, then the map pair (ψ, φ), is called an isomorphism between graphs X and Y. In other words, two graphs X and Y are isomorphic, if they have the same number of vertices and edges, and their vertices and edges are one-to-one mapped respectively, and the vertex-edge incidence relationships are kept to be constant (Fig. 7). Automorphism refers to the isomorphism of a graph to itself.

Clearly, if two graphs are isomorphic, the degree of their corresponding vertices is the same.

Given that $V(X) = V(Y)$, $E(X) = E(Y)$, $\varphi_X = \varphi_Y$, then the graphs X and Y are identical. Two identical graphs can be represented by a graph.

Divide the cut vertices of each graph in X and Y into two sets respectively, so that X and Y become two unconnected graphs. If the resulting two unconnected graphs are isomorphic, the two graphs X and Y are called 1-isomorphic. If a graph is a separable graph, it may contain 1-isomorphic graph.

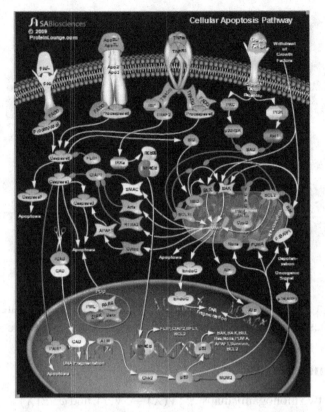

Figure 16.   A cell apoptosis pathway (SABiosciences, 2009).

3.   Biological Network Graphs

3.1.   *Graphs for species interactions*

In a community, a number of species form an ecological interaction graph through predation, parasitism, competition, symbiosis, etc., as indicated in Fig. 8. Obviously, food web is a kind of interaction graph, which is one of the most common natural phenomena.

3.2.   *Ecosystem graph*

In an ecosystem, species interact with each other and the environment, which forms a graph. An ecosystem graph may contain several vertices, as shown in Fig. 9, or tens of thousands of vertices.

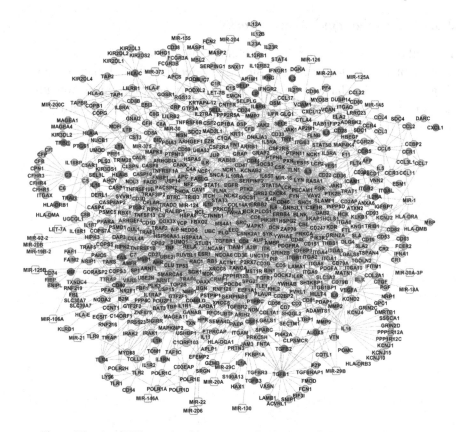

Figure 17. A miRNA-regulated immunoregulatory network (Tacutu *et al.*, 2011).

Ecosystem graph can be used to describe the relationship between various factors, which is a conceptual framework, as shown in Fig. 10.

3.3. *Eco-tourism network*

Ecological landscapes, tourist spots, roads, etc., can join together to generate a network and form an eco-tourism network (Fig. 11).

3.4. *Pest spread network*

Pests spread from one place to another place and thus generate various paths. It forms a spread path network (Fig. 12).

3.5.   *Metabolic regulatory networks*

In a cell or organelle, a variety of proteins, lipids, minerals, etc., can form a metabolic regulatory network through material flow and energy flow (Figs. 13–14). Similarly, there are gene regulatory networks, immunoregulatory networks, etc. (Figs. 15–17). Gene and metabolic regulatory networks are one of the science frontiers.

Fundamentals of Topology

Graph theory is a branch of topology. The term 'topology' was first introduced in 1847 by Professor JB Listing, a student of Professor Carl Friedrich Gauss. Topology mainly discusses the invariants and invariable properties of space in the topological transformation (homeomorphism). The early problems of graph theory, for example, Four-color Problem, have directly contributed to the birth and development of topology (Gross and Yellen, 2005). Therefore, the theories and methods of topology (Spanier, 1966; Chan, 1987; Lin, 1998) is an important basis of graph theory.

1. Topological Space, Subspace

1.1. *Topological space*

Let X be a set, Γ is a family of subsets of X, whose members satisfy the open set axiom:

(1) Both X and the empty set ϕ belong to Γ.
(2) If $O_1, O_2 \in \Gamma$, then $O_1 \cap O_2 \in \Gamma$.
(3) For any number of members in Γ, their union belongs to Γ.

Γ is called a topology on X, and any member of Γ is called the open set of X. The set X together with its topology Γ is called a topological space, denoted by (X, Γ). Known the topology Γ, the topological space (X, Γ) is denoted by X.

The topology consisting of X and the empty set ϕ is the trivial topology on X. The topology consisting of all subsets of X is called the discrete topology.

21

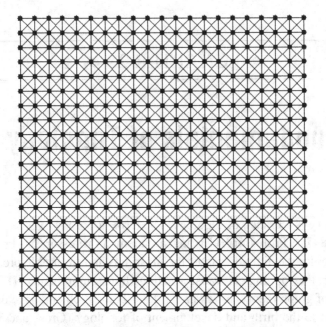

Figure 1.   A topological space of graph.

For the graph X, its elements are vertices and edges. A family of sub-graphs is Γ, which satisfies the open set axiom. Clearly, the subgraph family, Γ, is a topology on graph X. Graph X, together with the topology Γ, form a topological space (X, Γ). For example, in Fig. 1, all possible squares and their interior, form a family of subgraphs, and the subgraph family is a topology of Fig. 1.

That subset family of X, $\Phi = \{\Phi_\alpha\}$, is defined as the topological basis of X, if

(1) $X = \underset{\alpha}{\cup} \Phi_\alpha$.

(2) If $x \in \Phi_\alpha \cap \Phi_\beta$, then there exists $\Phi_\gamma \in \Phi$, such that $x \in \Phi_\gamma \subset \Phi_\alpha \cap \Phi_\beta$.

For example, in Fig. 1, the above-mentioned subgraph family is a topological basis.

Let $\Phi = \{\Phi_\alpha\}$ be the topological basis of X, then the subset A of X, is called the open set relative to Φ, if A is the union of some members of Φ.

The subset A of X is an open set relative to the topological basis Φ, if and only if, whenever $a \in A$, there exists $\Phi_\alpha \in \Phi$, such that $a \in \Phi_\alpha \subset A$.

Theorem 1. *If* $\Phi = \{\Phi_\alpha\}$ *is the topological basis of X, then the subset family* Γ, *consisting of open sets relative to* Φ, *is a topology of X, and* $\Phi \subset \Gamma$.

Clearly, the topological basis Φ is also a topology of X, and Γ is the Φ-induced topology. Two topological bases are equivalent if the topologies induced by them are the same.

Theorem 2. *Two topological bases,* Φ *and* Φ' *are equivalent if and only if each member of* Φ *is the open set relative to* Φ', *and each member of* Φ' *is the open set relative to* Φ.

In Fig. 1, the above topological basis is equivalent to a family of topological bases of subgraphs which consists of all squares and their interior from top left to bottom right corner in Fig. 1.

1.2. *Neighborhood, closed set, convergence of point sequence, and covering*

Any of open sets U of a topological space X is called the neighborhood of each $x \in U$. It is also the neighborhood of its subset $A \subset U$.

If A is a subset of topological space X, then $a \in A$ is called the interior point of A in X, if a has a neighborhood $U \subset A$. All interior points of A in X is called the interior of A in X, denoted by Int A.

The subset A of a topological space X is called the closed set of X, if the complimentary set of A, X-A, is an open set of X.

Closed Set Theorem. Let Ω be a family consisting of all closed sets of a topological space X, then

(1) X and the empty set ϕ belong to Ω.
(2) If $O_1, O_2 \in \Omega$, then $O_1 \cup O_2 \in \Omega$.
(3) For any number of members in Ω, their intersection set belongs to Ω.

Let A be a subset of a topological space X, and $x \in X$. If the intersection set of any neighborhood $B(x, \varepsilon)$ of x and A-$\{x\}$ is nonempty, then x is called the accumulation point of A. The union of A and all of its accumulation points is called the closure of A, denoted by $\bar{A}$. A is called the dense subset of X, if $\bar{A} = X$.

Let $\{x_n\}$ be a point sequence of the topological space X, and $a \in X$. For each neighborhood of a, $B(a, \varepsilon)$, if there is a natural number N, so that for all $n > N$, $x_n \in B(a, \varepsilon)$, then $\{x_n\}$ is said to converge to the point a, denoted by $\{x_n\} \to a$.

A is an open set of a topological space X, if and only if $A = \text{Int}A$. A is a closed set of X, if and only if $A = \bar{A}$.

Let A be a subset of a topological space X, and $\Theta = \{\Theta_\alpha\}$ is a subset family of X, such that

$$A \subset \bigcup_\alpha \Theta_\alpha,$$

Θ is called the covering of A in X. For each $\Theta_\alpha \in \Theta$, if Θ_α is an open set of X, Θ is called open covering of A in X. If Θ has only finite number of members, then Θ is called the finite covering of A in X. If the subset family $\Theta_0 \subset \Theta$, then Θ_0 is called the sub-covering of Θ.

For a graph, if A is a subgraph, a covering can thus be constructed.

1.3. Subspace, compact space

Let (X, Γ) be a topological space, A is a nonempty set of X, then the subset family $\Gamma|_A = \{O \cap A | O \in \Gamma\}$ is a topology of A, which is defined as the induced topology of Γ on A. The topological space $(A, \Gamma|_A)$ is called the subspace of the topological space (X, Γ).

For example, in Fig. 1, we can take an arbitrary subgraph, and construct a subspace according to the method above.

For any open covering Θ, of a topological space X, if there exists a finite sub-covering Θ_0, then X is called compact topological space. For the graph, we can also define compact space and it can be defined as a compact map.

1.4. Product space

Let $X_1, X_2, \ldots,$ and X_n be topological spaces, and $X = X_1 \times X_2 \times \cdots \times X_n$. The subset family of X, $U = \{U_1, U_2, \ldots, U_n | U_i$ is the open set of $X_i\}$, is a topological basis. The topology Γ, induced by this topological basis is called topological product on X. The topological space (X, Γ) is called the product space of $X_1, X_2, \ldots, X_n$, and X_i is called the ith factor space of X.

For a number of graphs, their composite graph can thus be constructed.

2. Continuous Map, Homeomorphism

Let $f: X \to Y$ be a single-valued correspondence from topological spaces X to Y, and $x_0 \in X$. For any neighborhood V of $f(x_0) \in Y$, if there exists a neighborhood U of x_0, such that $f(U) \subset V$, f is called to be continuous at x_0. If f is continuous at each point of X, f is called the continuous map from X to Y, sometimes called map (or mapping).

If $f: X \to Y$, and $g: Y \to Z$ are maps, then the composite map, $g \bullet f: X \to Z$, is also a map, of which, $(g \bullet f)(x) = g(f(x))$, $x \in X$.

The restriction of the map $f: X \to Y$ to the subspace A of X, $f|_A: A \to Y$, is $f|_A(a) = f(a)$, whenever $a \in A$. On the contrary, f is called the extension of $f|_A$.

Theorem 3. *Let f be the single-valued correspondence between topological spaces X and Y. The following statements are equivalent:*

(1) *f is a map from X to Y.*
(2) *For any open set V of Y, $f^{-1}(V)$ is an open set of X.*
(3) *For any closed set U of Y, $f^{-1}(U)$ is a closed set of X.*
(4) *For any subset A of X, $f(\bar{A}) \subset$ the closure of $f(A)$.*

If the map $f: X \to Y$ maps each open set of X to the open set of Y, then f is called an open map.

The map $f: X \to Y$ such that

$$f: X \cong f(X),$$

then f is called the embedded map.

Theorem 4. *Let $f: X \to Y$ be the single-valued correspondence between topological spaces X and Y. If f is continuous at x, then $\{x_n\} \to x \Rightarrow \{f(x_n)\} \to f(x)$.*

Let $f: X \to Y$ be a monogamy map, and its inverse $f^{-1}: Y \to X$ is also a map. f is called homeomorphism of X to Y, or topological transformation, denoted by

$$f: X \cong Y.$$

If there is such a homeomorphism f, we call topological spaces X and Y are homeomorphic, or topologically equivalence, denoted by $X \cong Y$.

f is a homeomorphism if and only if f is both injective and is onto open map.

The properties being invariant under homeomorphism (topological transformation) is called topologically invariant properties. Similarly, we have topological invariants. Open set, interior, closure, the convergence of point sequence, etc., are topologically invariant properties. For example, if f is a homeomorphism of X to Y, then U is an open set of X if and only if $f(U)$ is an open set of Y.

Homeomorphism of topological spaces is similar to isomorphism of graphs.

3.   Homotopy, Homotopy Type

Homeomorphism, or topologically invariant properties, is in essence homeomorphism classification. It is usually hard to treat these issues. However, the space classification under homotopy equivalence is relatively easy.

Two maps f_0, f_1: $X \to Y$ are called homotopic, if there exists a continuous map F: $X \times I \to Y$, such that

$$F(x, 0) = f_0(x), \quad F(x, 1) = f_1(x), \quad \forall x \in X.$$

F is called the homotopy from f_0 to f_1, denoted by $f_0 \simeq f_1$.

In general, for $f \simeq g$: $X \to Y$, we can construct $F(x, t) = (1 - t) f(x) + t g(x)$, $x \in X, t \in I$. In addition, $f \simeq g$: $X \to Y$, is equivalent for any $x \in X$, $f(x)$ can be in Y continuously deformed to $g(x)$.

In the set consisting of all maps from X to Y, the homotopy relation, denoted by $\simeq$, is an equivalence relation from X to Y.

All maps from X to Y can be classified into some equivalence categories based on homotopy relation, i.e., homotopy categories. The homotopy category of f is denoted by $[f]$.

Given that $f_0 \simeq f_1$: $X \to Y$, and f_1 is a constant map, i.e., $f_1(X) = a \in Y$, f_0 is called a null homotopy.

Theorem 5. *Let Y be a subspace of the Euclidean space R^n and f, g: $X \to Y$ are maps. $f \simeq g$, if for each $x \in X$, $f(x)$ and $g(x)$ can be connected by a line segment in Y.*

Let A be a subspace of X, the ordered pair (X, A) is called space pair. If the map f: $X \to Y$, maps the subspace A of X onto the subspace B of Y, f is called the map between space pairs, denoted by f: $(X, A) \to (Y, B)$.

Two maps between space pairs, f, g: $(X, A) \rightarrow (Y, B)$, are called to be homotopic, if there exists a map between space pairs F: $(X \times I, A \times I) \rightarrow (Y, B)$, such that $F(x, 0) = f(x)$, $F(x, 1) = g(x)$, $x \in X$.

Let f, g: $(X, A) \rightarrow (Y, B)$ be the maps between space pairs, such that $f|_A = g|_A$. f and g are called homotopic relative to A, if there exists a map between space pairs, F: $(X \times I, A \times I) \rightarrow (Y, B)$, such that $F(x, 0) = f(x)$, $F(x, 1) = g(x)$, $x \in X$, and $F(a, t) = f(a) = g(a)$, $a \in A, t \in I$, denoted by $f \simeq g$ rel A.

The topological spaces X and Y are called homotopy equivalent, if there exist maps f: $X \rightarrow Y$ and g: $Y \rightarrow Z$, such that $gf \simeq 1_X$, $fg \simeq 1_Y$, denoted by $X \simeq Y$.

Theorem 6. *The homotopy equivalence between topological spaces is an equivalence relation.*

The subspace A of X is called the retract of X, if there exists a map r: $X \rightarrow A$, such that $r(a) = a$, $\forall a \in A$. Let i: $A \rightarrow X$ be an interior map, i.e., $i(a) = a$, $\forall a \in A$, then $ri = 1_A$. The map r is called retraction map. If there is a homotopy $ir \simeq 1_X$, then the homotopy is called deformation retraction, and A is the deformation retract of X.

If A is the deformation retract of X, then $A \simeq X$, in which the interior map, i: $A \rightarrow X$ is a homotopy equivalence, and r: $X \rightarrow A$ is the homotopy inverse of i.

4. Connectedness

4.1. *Connectedness*

Let X be the union of the disjoint nonempty sets A and B of X, X is thus called unconnected space, otherwise connected space.

The continuous image of a connected space is a connected set. Connectedness of topological space is a topologically invariant property.

The subset A of a topological space X is called the connected subset of X, if A, as a subspace, is connected.

Theorem 7. *Suppose the topological space X has a covering Θ, which consists of connected subsets, such that for any two members A and B in Θ, there exists finite number of members of Θ, $A = \Theta_1, \Theta_2, \ldots, \Theta_n = B$, such that $\Theta_i \cap \Theta_{i+1} \neq \phi$, $i = 1, 2, \ldots, n-1$, then X is connected.*

Let A be a connected subset of X and not be a proper subset of other connected sets, A is called the connected component of X. The connected component of X is a closed set. Different connected components are disjoint. X is the union of all of its connected components.

Topological space X is locally connected, if for any $x \in X$ and the neighborhood U_x of x, there exists a connected neighborhood V_x of x, such that $V_x \subset U_x$. Topological space X is locally connected if and only if the connected component of any open set of X is the open set. Local connectedness is a topologically invariant property.

4.2.   *Path connectedness*

The map $f: I \to X$ that satisfies $f(0) = a$, $f(1) = b$, is called a path connecting a and b in X. X is said to be path connected if for any points a and b in the topological space X, there exists a path connecting them. a is called the origin of the path and b the end.

Path connectedness of topological space corresponds to the connectedness of graph.

Path connectedness is a topological invariant property. The continuous image of path connected space is path connected.

The subset A of a topological space X is called path connected subset, if A is, as a subspace, path connected. If A is a path connected subset of X, but not a proper subset of other path connected sets, then A is called path connected component. For the graph, the path connected subset is the connected subgraph.

Every topological space is the union of its disjoint path connected components.

A path connected space must be connected space. However, a connected space is not necessarily path connected space. Taking a graph as a topological space, then Theorem 7 holds also for graph.

Corollary 1. *Let the graph X be a topological space, there exists a covering Θ, consisting of connected subgraphs, such that for any two members A and B in Θ, there exist a finite number of members of Θ, $A = \Theta_1, \Theta_2, \ldots, \Theta_n = B$, such that $\Theta_i \cap \Theta_{i+1} \neq \phi$, $i = 1, 2, \ldots, n-1$, then the graph X is a connected graph.*

Topological space X is locally path connected, if for any $x \in X$, and the neighborhood U_x of x, there exists a path connected neighborhood V_x of x, such that $V_x \subset U_x$. Locally path connected space is also locally connected space.

Topological space X is locally connected if and only if the connected component of any open set of X is the open set. Local connectedness is a topologically invariant property.

$X_1 \times X_2 \times \cdots \times X_n$ is connected (path connected, locally connected, locally path connected), if and only if each X_i is connected (path connected, locally connected, locally path connected).

5. Simplicial Complex, Polyhedron

5.1. *Simplex*

The $n + 1$ points, $a^0, a^1, \ldots$, and a^n in Euclidean space R^m are said to be geometry-independent, if the vectors $a^1\text{-}a^0, a^2\text{-}a^0, \ldots, a^n\text{-}a^0$ are linearly independent.

Let $a^0, a^1, \ldots, a^n$, be geometry-independent point set in Euclidean space R^m. The set

$$\sigma_n = \left\{ \sum_{i=0}^{n} \lambda_i a^i \,|\, \lambda_i \geqslant 0, \sum_{i=0}^{n} \lambda_i = 1 \right\}$$

is given the subspace topology of R^m. It is called n-dimensional simplex, denoted by $(a^0, a^1, \ldots, a^n)$, or σ_n. $a^0, a^1, \ldots$, and a^n are called vertices of n-dimensional simplex. Clearly, the point x of σ_n can be uniquely expressed as

$$x = \sum_{i=0}^{n} \lambda_i a^i.$$

If $\lambda_i > 0$, $i = 1, 2, \ldots, n$, then x is called the interior of σ_n, otherwise boundary point. The simplex extended from the subset of the vertex set $\{a^0, a^1, \ldots, a^n\}$ is called the facet of σ_n.

Zero-dimensional simplex is a point, Two-dimensional simplex is a line segment, Three-dimensional simplex is a triangle, and Four-dimensional simplex is a tetrahedron, etc.

The simplex $\sigma_n = (a^0, a^1, \ldots, a^n)$ is the minimum convex set that contains the vertices $a^0, a^1, \ldots,$ and a^n, and it is a compact, closed and connected space in R^m.

Any two 2-dimensional simplexes are linearly homeomorphic.

5.2. Simplicial complex

The simplicial complex K is a set consisting of a finite number of simplexes in R^m, and satisfies the conditions:

(1) If $\sigma_n \in K$, then any facet of $\sigma_n \in K$.
(2) If $\sigma_n, \tau_m \in K$, then $\sigma_n \cap \tau_m$ is either empty or a shared facet.

The maximum dimension of simplexes in K is called the dimension of the complex K (dim K). Each Zero-dimensional simplex of K is called the vertex of K. For example, the complex $K = \{(a^0 a^1 a^2), (a^0 a^1),$ $(a^1 a^2), (a^0 a^2), (a^0 a^3), (a^1 a^4), (a^0), (a^1), (a^2), (a^3), (a^4)\}$, dim $K = 2$, a^0, $a^1, a^2, a^3,$ and a^4, are the vertices of K (Fig. 2).

The subset L of K is called the subcomplex of K, if L satisfies the condition (1) (and naturally (2)). (K, L) is a simplex pair. If L and M are subcomplexes of K, then $L \cap M$ and $L \cup M$ are also the subcomplexes of K.

5.3. Polyhedron

The union of all simplexes of the complex K

$$|K| = \bigcup_{\sigma \in K} \sigma$$

is given the subspace topology of R^m, which is called polyhedron of the simplex K. The polyhedron $|L|$ of subcomplex L of K is called subpolyhedron of the polyhedron $|K|$.

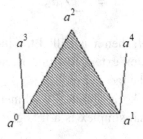

Figure 2. Complex K.

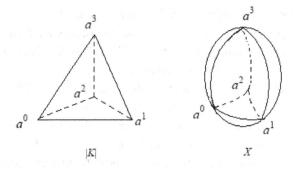

Figure 3. Simplex partition of topological space X.

The polyhedron $|K|$ of complex K is a compact and closed subspace of R^m.

(K, f) is called a simplex partition of a topological space X (Fig. 3), if K is a simplicial complex, and $f: |K| \to X$ is a homeomorphism. Topological space X is called polytope, or curved polyhedron.

5.4. *Abstract complex, graph*

Abstract complex $\mathcal{K}$ is a family consisting of a set of a finite number of elements (abstract vertices), $a^0, a^1, \ldots, a^n$, together with the subsets $(a^{i0}, a^{i1}, \ldots, a^{ir})$ (i.e., abstract simplexes), such that every subset that satisfies every abstract simplex is still abstract simplex.

One-dimensional abstract complex $\mathcal{K}$ is called graph.

If there exists an one-to-one correspondence $f: K \to \mathcal{K}$ between geometric complex K and abstract complex $\mathcal{K}$, such that $(a^{i0}, a^{i1}, \ldots, a^{ir})$ is the simplex of K, if and only if $(f(a^{i0}), f(a^{i1}), \ldots, f(a^{ir}))$ is the abstract simplex of $\mathcal{K}$. In this case, K is called the geometric realization of $\mathcal{K}$.

n-dimensional abstract complex $\mathcal{K}$ can be realized in Euclidean space R^{2n+1} but not necessarily in R^{2n}. So whether a graph (one-dimensional abstract complex) is planary (i.e., planar graph; R^2) or not, is conditional, which is called the planary condition. A planary graph is called the planar graph, or else the nonplanar graph.

Theorem 8. *Let geometric complexes K_1, K_2 be geometric realization of the abstract complex $\mathcal{K}$ achieved, then there exists a homeomorphism $f: |K_1| \to |K_2|$, which maps simplex into the simplex with the same dimension.*

Suppose the abstract complexes $\mathcal{K}$ and $\mathcal{L}$ have vertices $a^0, a^1, \ldots$, and $b^0, b^1, \ldots$, respectively, and the geometric complexes K, L are the geometric realization of $\mathcal{K}$ and $\mathcal{L}$ respectively. The union $\mathcal{K}*\mathcal{L}$ is defined as the abstract complex with the abstract vertices $a^0, a^1, \ldots, b^0, b^1, \ldots$, in which the abstract simplexes are those that the subsets $(a^{i0}, a^{i1}, \ldots, b^{j0}, b^{j1}, \ldots,)$ such that $(a^{i0}, a^{i1}, \ldots)$ are the abstract simplexes of $\mathcal{K}$, and $(b^{j0}, b^{j1}, \ldots)$ are the abstract simplexes of $\mathcal{L}$. Any geometric realization of $\mathcal{K}*\mathcal{L}$ is called the union of geometric complexes K and L, denoted by $K*L$.

The union of two or more complexes can be defined in the same fashion, and $(K*L)*M = K*(L*M)$, denoted by $K*L*M$. The partition of spherical surface S^{n-1} (in Fig. 3, $n = 4$) can be written as $K_1*K_2*\cdots*K_n$ (in Fig. 3, it is $K_1*K_2*K_3*K_4$), in which K_i is the simplicial complex consisting of two zero-dimensional simplexes (points).

K is a complex in R^m, and L is a complex in R^n, then $K*L$ can be in general realized in R^{m+n+1}.

5.5.   *Connectedness of complex*

The complex K is said to be connected, if it is not the union of two nonempty and disjoint subcomplexes. L is called a connected component of K, if the subcomplex L of the complex K is connected and L is not a proper subcomplex of another connected subcomplex of K.

Complex K is connected, if and only if for any two vertices a and b of K, there exist a sequence of vertices of K, $a = a^0, a^1, \ldots, a^{n-1}, a^n = b$, such that $(a^i, a^{i+1}), i = 1, 2, \ldots, n-1$, are all one-dimensional simplexes of K.

The connectedness of complex corresponds to the connectedness of graph.

Any complex K can be decomposed into r connected components $K_1, K_2, \ldots, K_r$, and K is the union of these disjoint connected components. The connected components of polyhedron $|K|$ are $|K_1|, |K_2|, \ldots, |K_r|$, and $|K|$ is the union of $|K_1|, |K_2|, \ldots, |K_r|$, where $K_1, K_2, \ldots$, and K_r are the connected components of K.

Theorem 9.   *Complex K is connected $\Leftrightarrow$ polyhedron $|K|$ is connected $\Leftrightarrow$ $|K|$ is path connected.*

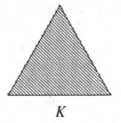 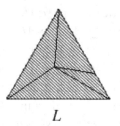

Figure 4.   Re-partition of complex K.

From Theorem 9, for the graph, connectedness is equivalent to path connectedness.

5.6.   *Re-partition, simplicial approximation*

The complex K is achieved by splitting the polyhedron $|K|$ into several small pieces. It is the simplicial partition of the polyhedron $|K|$.

Complex L is called the re-partition (Fig. 4) of complex K, if $|L| = |K|$, and each simplex of L is contained in a simplex of K.

Known complexes K and L, the correspondence $f: |K| \rightarrow |L|$ is called a simplicial map, if it satisfies the conditions:

(1) For each vertex a of K, $f(a)$ is the vertex of L.
(2) Given that $(a^0, a^1, \ldots, a^n)$ is a simplex of K, then $f(a^0), f(a^1), \ldots,$ $f(a^n)$ can be extended into a simplex of L.
(3) If

$$\lambda = \sum_{i=0}^{n} \lambda_i a^i$$

is a point of the simplex $(a^0, a^1, \ldots, a^n)$ of K, then we have

$$f(x) = \sum_{i=0}^{n} \lambda_i f(a^i).$$

The simplicial map $f: |K| \rightarrow |L|$ is a continuous map.

Any point x of the polyhedron $|K|$ must belong to the interior of the unique simplex τ of K. τ is called the carrying simplex of x in K, denoted by $\tau = \mathrm{Car}_K(x)$.

The simplicial map, $\psi: |K| \to |L|$ is called the simplicial approxima-tion of the map, $f: |K| \to |L|$, if for any $x \in |K|$, $\psi(x) \in \mathrm{Car}_L f(x)$. In other words, for any $x \in |K|$, $\psi(x)$ and $f(x)$ are in the same simplex.

The simplicial map, $\psi: |K| \to |L|$, is the simplicial approximation of the map, $f: |K| \to |L|$, if and only if for each vertex a of K, $f\,(\mathrm{st}_K a) \subset \mathrm{st}_L \psi(a)$, where $\mathrm{st}_K a$ is the union of interiors of all simplexes τ with a as a vertex, which is an open subset of $|K|$ and called open star.

Theorem 10. *If $f: |K| \to |L|$ is a map, and for each vertex a of K, there exists the vertex b of L, such that $f(\mathrm{st}_K a) \subset \mathrm{st}_L b$, then f has a simplicial approximation, $\psi: |K| \to |L|$, such that $\psi(a) = b$.*

Theorem 11.
(1) *If the simplicial maps ψ_1, ψ_2 are simplicial approximation of the maps $f_1: |K| \to |L|$, $f_2: |L| \to |M|$, then $\psi_2 \psi_1$ is the simplicial approxi-mation of $f_2 f_1$.*
(2) *If $f: |K| \to |L|$ has a simplicial approximation, $\psi: |K| \to |L|$, then $f \simeq \psi \operatorname{rel} A$, where $A = \{x \in |K| | f(x) = \psi(x)\}$.*

Theorem 12. *For any map $f: |K| \to |L|$, there exists a positive integer r, such that $f: |K^{(r)}| \to |L|$ has a simplicial approximation.*

6.   Fundamental Group

Fundamental group is homotopy equivalently invariant. It is a topologi-cal invariant. If a topological space X is a polytope, we may calculate the fundamental group according to simplicial approximation theorems (The-orems 10, 11, and 12).

6.1.   *Product path*

Let X be a topological space, and the points x, $y \in X$. A map $u: I \to X$ such that $u(0) = x, u(1) = y$, which is called a path (Fig. 5) from x to y in X. If $u(0) = u(1) = x_0$, it is called a closed path of X with x_0 as the base point.

For the polyhedron corresponding to a graph, the definition of path can be applied.

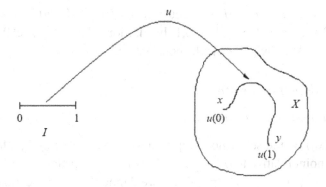

Figure 5. A path from x to y in topological space X.

Suppose the paths $u, v: I \to X$ such that $u(1) = v(0)$, the product path $u^*v: I \to X$ is defined by

$$u^*v(t) = u(2t), \quad 0 \leqslant t \leqslant 1/2;$$
$$v(2t-1), \quad 1/2 \leqslant t \leqslant 1.$$

Suppose the paths $u_1, u_2, \ldots, u_n: I \to X$ such that $u_i(1) = u_{i+1}(0)$, $i = 1, 2, \ldots, n-1$, the product path $u_1^*u_2^* \cdots {}^*u_n: I \to X$ is defined by

$$u_1^*u_2^* \cdots {}^*u_n(t) = u_i(nt-i+1), \quad (i-1)/n \leqslant t \leqslant i/n, \ i = 1, 2, \ldots, n$$

Known that the paths of X, $u_1, u_2, \ldots, u_n$, and $v_1, v_2, \ldots, v_n$, such that $u_i(0) = v_i(0)$, $u_i(1) = v_i(1)$, $i = 1, 2, \ldots, n$, and $u_i(1) = u_{i+1}(0)$, $v_i(1) = v_{i+1}(0)$, $i = 1, 2, \ldots, n-1$. If $u_i \simeq v_i$ rel $0, 1$, $i = 1, 2, \ldots, n$, then we have

$$u_1^*u_2^* \cdots {}^*u_n \simeq v_1^*v_2^* \cdots {}^*v_n \text{ rel } 0, 1,$$

and

$$(u_1^*u_2^* \cdots u_i)^*(u_{i+1}^* \cdots {}^*u_n) \simeq u_1^*u_2^* \cdots {}^*u_n \text{ rel } 0, 1,$$

i.e., product paths are associative in the sense of homotopy.

The inverse path, $u^{-1}: I \to X$, of the path u in X, is defined by $u^{-1}(t) = u(1-t)$, $0 \leqslant t \leqslant 1$, where u^{-1} is a path with the end of u as origin, and with the origin of u as end, and the trajectory is the same as u.

Known the paths, u, v: $I \to X$, such that $u(0) = v(0)$, $u(1) = v(1)$. If $u \simeq v$ rel $0,1$, then $u^{-1} \simeq v^{-1}$ rel $0,1$. In addition, $u*u^{-1} \simeq c_x$ rel $0,1$, and $u^{-1}*u \simeq c_y$ rel $0,1$, where c_x is the constant path.

6.2.   Fundamental group

6.2.1.   Fundamental group

Let $\pi_1(X, x_0)$ be the set of homotopy categories of the closed paths with x_0 as base point relative to $0,1$, i.e., $\pi_1(X, x_0) = \{[u]|$ path $u: I \to X$ such that $u(0) = u(1) = x_0\}$. On $\pi_1(X, x_0)$ we define the multiplication to be $[u] \times [v] = [u*v]$, then

(1) $([u] \times [v]) \times [w] = [u] \times ([v] \times [w])$.
(2) $[u] \times [u^{-1}] = [c_{x0}]$; $[u] \times [c_{x0}] = [c_{x0}] \times [u]$.

Therefore, $\pi_1(X, x_0)$ is a group, which is called the fundamental group of X with x_0 as base point.

Theorem 13. *A map $f: (X, x_0) \to (Y, y_0)$ induces a homomorphism*

$$f_*: \pi_1(X, x_0) \to \pi_1(Y, y_0)$$

which satisfies the conditions: (1) if $f': (X, x_0) \to (Y, y_0)$ such that $f \simeq f'$ rel x_0, then $f_* = f'_*$; (2) if $g: (Y, y_0) \to (Z, z_0)$, then $(gf)_* = g_* f_*$.

If $(X, x_0) \simeq (Y, y_0)$, then f induces a homomorphism, $f_*: \pi_1(X, x_0) \to \pi_1(Y, y_0)$.

The set of all path connected components of a topological space X is $\pi_0(X)$. If $X \simeq Y$, then there is an one-to-one correspondence relation between $\pi_0(X)$ and $\pi_0(Y)$.

Let x, y be any two points in a path connected space X, then $\pi_1(X, x) \cong \pi_1(X, y)$. In other words, fundamental group is independent of the choice of base point in the sense of homomorphism. Therefore, $\pi_1(X, x)$ can be denoted by $\pi_1(X)$.

Fundamental group of the connected graph is independent of the choice of base point.

Theorem 14. *Given $X \simeq Y$, then $f_*: \pi_1(X, x_0) \to \pi_1(Y, f(x_0))$ is a isomorphism.*

6.2.2. *Fundamental group of polyhedron*

The fundamental group of general topological spaces is hard to calculate. However, the fundamental group of polyhedron $X = |K|$ can be calculated according to simplicial approximation theorems.

The edge path of complex K, from vertices a^0 to a^n, is a sequence of vertices $a, a^1, \ldots, a^n$, such that all (a^{i-1}, a^i) are extended into one-dimensional or zero-dimensional simplexes of K, $1 \leqslant i \leqslant n$. If $a^0 = a^n$, then it is called a closed edge path with a^0 as the base point. If there is only one vertex a^0, it is called constant edge path.

For the graph, one-dimensional or zero-dimensional simplex corresponds to edge and vertex, edge path corresponds to chain (edge sequence), and closed edge path corresponds to closed chain.

The product edge path of two edge paths, $\alpha = a^0 a^1 \cdots a^n$, $\beta = a^n a^{n+1} \cdots a^{n+m}$, is $\alpha \bullet \beta = a^0 a^1 \cdots a^n a^{n+1} \cdots a^{n+m}$. The inverse edge path of α is $\alpha^{-1} = a^n \cdots a^1 a^0$.

Admissible transformation of edge path is defined as:

(1) if $a^{r-1} = a^r$, the edge path $\ldots a^{r-1} a^r \ldots$ becomes $\ldots a^r \ldots$, and vice versa;
(2) if a^{r-1}, a^r, a^{r+1} are extended into the simplexes of K, the edge path $\ldots a^{r-1} a^r a^{r+1} \ldots$ becomes $\ldots a^{r-1} a^{r+1} \ldots$, and vice versa.

If edge path α becomes β through a finite number of admissible transformation, then α is said to be equivalent to β, denoted by $\alpha \sim \beta$.

If α_0, β_0 are edge paths from vertices a^0 to a^n, and α_1, β_1 are edge paths from vertices a^n to a^{n+m}, and $\alpha_0 \sim \beta_0, \alpha_1 \sim \beta_1$, then

(1) $\alpha_0 \bullet \alpha_1 \sim \beta_0 \bullet \beta_1$;
(2) $\alpha_0^{-1} \sim \beta_0^{-1}$;
(3) $\alpha_0 \bullet \alpha_0^{-1} \sim a^0, a^0 \bullet \alpha_0 \sim \alpha_0 \sim \alpha_0 \bullet a^n$.

Suppose $\pi(K, a^0)$ is the set of equivalence categories of closed edge paths with a^0 as the base point in complex K. In this set, define multiplication as $[\alpha] \bullet [\beta] = [\alpha \bullet \beta]$, then $\pi(K, a^0)$ is an edge path group of complex K, with a^0 as the base point.

Theorem 15. $\pi(K, a^0) \cong \pi(|K|, a^0)$.

Therefore, the fundamental group of polyhedron is equivalent to the edge path group of complex.

One-dimensional subcomplex L of complex K is called tree, if $|L|$ is contractible, and vice versa. The tree L of complex K is called maximal tree if there is not a tree L' such that $L' \subset L$. If $|K|$ is path connected and L is the maximal tree of K, then L contains all vertices of K. The tree of a one-dimensional connected complex with v vertices contains $v - 1$ one-dimensional simplexes (a^{i-1}, a^i). These properties are the same to the tree of graph.

Suppose $|K|$ is path connected. Rank all vertices of K as $a^0 < a^1 < \cdots < a^n$, and each simplex of K is denoted by $(a^{i0}, a^{i1}, \ldots, a^{ir})$, where $a^{i0} < a^{i1} < \cdots < a^{ir}$, which is called ordered simplex. Each ordered simplex of K is $g_{ij} = (a^i, a^j)$. In addition, let $|L|$ be a contractible polyhedron containing all vertices of K. Define G as a group, and its generators system and relations system are

Generators system: $\{g_{ij} | (a^i, a^j)$ is the one-dimensional ordered simplex of $K \backslash L\}$.

Relations system: $\{g_{ij} g_{jk} g_{ik}^{-1} | (a^i, a^j, a^k)$ is the two-dimensional ordered simplex of $K \backslash L\}$

$$\{g_{ij} = 1 | (a^i, a^j) \in L\}.$$

Theorem 16. $G \cong \pi(K, a^0)$.

Theorem 17. $\pi_1(X \times Y, (x_0, y_0)) \cong \pi_1(X, x_0) \times \pi_1(Y, y_0)$.

The fundamental group of polyhedron is a finitely generated group.

CHAPTER 3

Matrix Representations and Computer Storage of Graphs

To analyze and calculate a graph, it is necessary to present a matrix representation of the graph. The matrix representation contains the topological information and properties of the graph (Chan *et al.*, 1982; Li, 1982; Gross and Yellen, 2005). In addition, a proper computer storage method is indispensible for a graph. We need a small storage space and easy access of the graph.

1. Undirected Graph

1.1. *Incidence matrix*

Incidence matrix describes the vertex-edge relationship of graph.

An undirected graph X with v vertices and e edges, has an incidence matrix $A = (a_{ij})_{v \times e}$, where $a_{ij} = 1$, if $(v_i, v_k) = e_j$, and $a_{ij} = 0$, if $(v_i, v_k) \neq e_j$; $i = 1, 2, \ldots, v$; $j = 1, 2, \ldots, e$.

For example, the incidence matrix A of the graph in Fig. 2(a) of Chapter 1, is as the following

$$A = (a_{ij})_{5 \times 10} = \begin{bmatrix} 1 & 0 & 0 & 0 & 1 & 1 & 1 & 0 & 0 & 0 \\ 1 & 1 & 1 & 0 & 0 & 0 & 0 & 0 & 0 & 1 \\ 0 & 0 & 1 & 1 & 0 & 0 & 0 & 1 & 0 & 0 \\ 0 & 0 & 0 & 1 & 1 & 1 & 0 & 0 & 1 & 0 \\ 0 & 0 & 0 & 0 & 0 & 0 & 1 & 1 & 1 & 1 \end{bmatrix}.$$

In the incidence matrix, the number of one's in a row is the degree of the vertex corresponding to that row. There are two one's in a column, which indicates the two vertices associated with the edge corresponding to that column.

For a connected graph, each row of an incidence matrix contains at least a single one. If there is only a single one in some row, then the edge of the column containing that single one is a suspended edge which does not appear in any circuit.

Suppose there is an incidence matrix A of the following type

$$A = \begin{bmatrix} A_{11} & 0 \\ 0 & A_{22} \end{bmatrix}.$$

then the graph is called an unconnected graph X, and it has at least two maximal connected subgraphs X_1 and X_2. For example, the incidence matrix of the unconnected graph in Fig. 1 is

$$\begin{bmatrix}
1 & 0 & 0 & 1 & 1 & 0 & 0 & 0 & 0 & 0 & 0 \\
1 & 1 & 0 & 0 & 0 & 0 & 0 & 1 & 0 & 0 & 0 \\
0 & 1 & 1 & 0 & 0 & 1 & 0 & 0 & 0 & 0 & 0 \\
0 & 0 & 1 & 1 & 0 & 0 & 1 & 0 & 0 & 0 & 0 \\
0 & 0 & 0 & 0 & 1 & 1 & 1 & 1 & 0 & 0 & 0 \\
0 & 0 & 0 & 0 & 0 & 0 & 0 & 0 & 1 & 1 & 1 \\
0 & 0 & 0 & 0 & 0 & 0 & 0 & 0 & 1 & 1 & 1
\end{bmatrix}.$$

The rank of the incidence matrix of a connected graph with v vertices is $v - 1$. For a graph with v vertices and p maximal connected subgraphs, the rank of its incidence matrix is $v - p$.

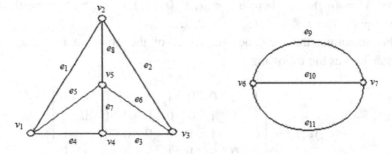

Figure 1.   An unconnected graph $X(X_1, X_2)$.

By removing the row corresponding to the vertex v_i from the incidence matrix of a connected graph X with v vertices and e edges, we obtain a $(v-1) \times e$ matrix, which is called the fundamental incidence matrix that corresponds to the vertex v_i. The rank of fundamental incidence matrix is $v-1$.

If there is a circuit in a fundamental incidence matrix, then the column vectors corresponding to the edges in the circuit are linearly dependent.

The elementary transformation of matrix can be operated to incidence matrix using the additive rules $(0+0 = 0, 0+1 = 1, 1+0 = 1, 1+1 = 0)$ and multiplication rules $(0 \bullet 0 = 0, 0 \bullet 1 = 0, 1 \bullet 0 = 0, 1 \bullet 1 = 1)$. By deleting a row from the incidence matrix, we can obtain an incidence submatrix.

A graph corresponds to a unique incidence matrix. In the sense of homomorphism, an incidence matrix corresponds to a unique graph.

1.2. Circuit matrix

Circuit matrix describes the edge-circuit relationship of graph.

An undirected graph X with v vertices, e edges, and c circuits, has a circuit matrix $C = (c_{ij})_{c \times e}$, where $c_{ij} = 1$, if the edge e_j is in the circuit c_i, and $c_{ij} = 0$, if the edge e_j is not in the circuit c_i; $i = 1, 2, \ldots, c$; $j = 1, 2, \ldots, e$.

For example, the circuit matrix of the graph in Fig. 2 is

$$C = (c_{ij})_{7 \times 6} = \begin{bmatrix} 1 & 0 & 0 & 1 & 0 & 1 \\ 0 & 1 & 0 & 0 & 1 & 1 \\ 0 & 0 & 1 & 1 & 1 & 0 \\ 1 & 1 & 1 & 0 & 0 & 0 \\ 1 & 1 & 0 & 1 & 1 & 0 \\ 1 & 0 & 1 & 0 & 1 & 1 \\ 0 & 1 & 1 & 1 & 0 & 1 \end{bmatrix}.$$

An undirected and connected graph X with v vertices and e edges, has a circuit matrix, of rank $e - v + 1$. Therefore, there are $e - v + 1$ circuits in the undirected and connected graph X.

Suppose A and C are incidence matrix and circuit matrix of an undirected and connected graph X, respectively, and if the edges represented by columns of two matrices are the same, then $AC^T = 0$, $CA^T = 0$.

For an undirected and connected graph X with v vertices and e edges, the fundamental circuit matrix of its tree T is $C = (c_{ij})_{(e-v+1)\times e}$, in which the rows are $e - v + 1$ fundamental circuits of the tree and the columns are e circuit branches. If the edge e_j is in the fundamental circuit c_i, then $c_{ij} = 1$ and if the edge e_j is not in the fundamental circuit c_i then $c_{ij} = 0$; where $i = 1, 2, \ldots, e - v + 1$; $j = 1, 2, \ldots, e$. The rank of fundamental circuit matrix is $e - v + 1$.

1.2.1. *Paton's fundamental circuit finding algorithm*

In a number of research on graphs, we need to calculate fundamental circuit set. The following is the algorithm of fundamental circuit set developed by Paton (1969):

Suppose the vertex set of a graph X is $V = \{1, 2, \ldots, n\}$, the adjacency matrix is D, the set of the vertices already on the tree is T, and the set of the vertices to be tested is S. Let $1 \in T$, $S = V$, and the vertex 1 be the tree root, then:

(1) If $T \cap S = \phi$, terminate calculation.
(2) If $T \cap S \neq \phi$, choose a vertex in $T \cap S$.
(3) Sequentially test every edge associated with the vertex v; if there is no edge to be tested then remove v from S, and return to (1).
(4) If there exists an edge (v, w) to be tested, test whether the vertex w is in T or not.
(5) If $w \in T$, find out the edge (v, w) and the fundamental circuit generated by the unique path (in the tree) that links v and w; remove the edge (v, w) from the graph and return to (3).
(6) If $w \notin T$, add the edge (v, w) to the tree and the vertex w to T; remove the edge (v, w) from the graph and return to (3).

The following are Java codes, fundCircuit, for calculating fundamental circuit set:

```
//Calculate the set of fundamental circuits of the graph
//Matrix d[n][n] is the adjacency matrix D.
public class fundCircuit {
public static void main(String[] args){
int i,j,v;
if (args.length!=2)
```

```
System.out.println("You must input the name of table in the
database. For example, you may type the following in the command
window: java fundCircuit 21 fundcircuit, where fundcircuit is
the name of table, 21 means the number of vertice.");
v=Integer.valueOf(args[0]).intValue();
String tablename=args[1];
readDatabase readdata=new readDatabase ("dataBase",
tablename,v);
v=readdata.m;
int a[][]=new int[v+1][v+1];
for(i=1;i<=v;i++)
for(j=1;j<=v;j++)
a[i][j]=(Integer.valueOf(readdata.data[i][j])).intValue();
fundCircuit(v,a); }
public static void fundCircuit(int v, int d[][]) {
int i,j,m,r,w,a,t,its,lm,num;
int l[]=new int[v+1];
int vp[]=new int[v+1];
int ts[]=new int[v+1];
int circuit[]=new int[10000];
num=0;
for(i=1;i<=v;i++) l[i]= -1;
t=1;
loop: do {
its=1;
ts[1]=t;
l[t]=0;
do {
if (its==0) break;
r=ts[its];
lm=l[r]+1;
for(w=1;w<=v;w++) {
if (d[r][w]<=0) continue;
else if ((l[w]+1)==0) {
ts[its]=w;
its++;
vp[w]=r;
l[w]=lm;
d[r][w]=0;
d[w][r]=0;
continue; }
num++;
a=vp[w];
m=1;
circuit[1]=r;
j=r;
```

```
do {
j=vp[j];
m++;
circuit[m]=j;
if (j==a) break; }
while(v>0);
m++;
circuit[m]=w;
System.out.print("No. fundanmental circuit: ");
System.out.println(num);
System.out.print("Fundanmental circuit: ");
for(j=1;j<=m;j++)
System.out.print(circuit[j]+"->");
System.out.print(circuit[1]);
System.out.println("\n");
d[r][w]=0;
d[w][r]=0; }
its--; }
while(v>0);
for(t=t;t<=v;t++)
if (l[t]==-1) continue loop;
break; }
while (v>0); }
}
```

The computational complexity of Paton's algorithm on fundamental circuit set is $O(n^k)(2 < k < 3)$, and the memory requirement is n^2 (Chan *et al.*, 1982).

1.2.2. *Chan's circuit matrix algorithm*

For an undirected and connected graph X with v vertices and e edges, an algorithm to calculate the circuit matrix C of X is (Chan *et al.*, 1982):

(1) Arbitrarily choose a tree T from a graph X, and write out the fundamental circuit matrix C_a of tree T.

(2) Make all possible ring sum operations in C_a, and construct a new matrix C_b by adding original rows of C_a, together with the new rows generated from ring sum operations.

(3) Eliminate redundant rows (i.e., the rows of the circuits with disjoint edges) in C_b.

(4) Construct the circuit matrix C from the left rows of C_b.

The algorithm can also be used to determine the graph planarity.

The following are Java class (they should be improved further for practical use), redunDel, for eliminating redundant rows:

```
//Delete redundant rows.
/*n: number of fundamental circuits of graph; e: number of edges
of
   graph. nt: number of edges of tree T. Matrix cc[n][e] is the
   matrix of fundamental circuits, Ca. The resulted matrix
   Cb, i.e., c[1-n][1-e], will be obtained. */
   public class redunDel {
   public int c[][];
   public redunDel(int n, int e, int nt, int cc[][]) {
   private int i, j, k, nn, nm, nr;
   c=new int[n+1][e+1];
   private int t[][]=new int[n+1][nt+1];
   private int mc[]=new int[n+1];
   for(j=1;j<=e;j++)
   for(i=1;i<=n;i++) c[i][j]=cc[i][j];
   for(j=1;j<=nt;j++)
   for(i=1;i<=n;i++) t[i][j]=0;
   for(j=1;j<=e;j++) {
   if (c[1][j]==0) continue;
   nm=0;
   for(i=1;i<=n;i++) {
   if (c[i][j]==0) break;
   nm++; }
   for(k=1;k<=e;k++) {
   if ((c[1][k]==0) | (k==j)) continue;
   nn=0;
   for(i=1;i<=n;i++) {
   if (c[i][k]==0) break;
   nn++; }
   for(i=1;i<=n;i++) {
   t[i][1]=c[i][j];
   t[i][2]=c[i][k]; }
   mc=ringSum(n, nt, t);
   nr=0;
   for(i=1;i<=n;i++) {
   if (mc[i]==0) break;
   nr++; }
   if ((nm-nn)!=nr) continue;
   for(i=1;i<=n;i++) c[i][j]=0;
   break; } }
   }
```

```
//Calculate ring sum.
public int[] ringSum(int n, int nt, int t[][]) {
private int i, j, s, m;
private int mc[]=new int[n+1];
private int lc[]=new int[n+1];
for(i=1;i<=n;i++)
{mc[i]=0;
lc[i]=0; }
for(j=1;j<=nt;j++)
for(i=1;i<=n;i++) {
if (t[i][j]==0) continue;
s=t[i][j];
lc[s]++; }
j=1;
for(s=1;s<=n;s++) {
m=lc[s]/2*2;
if (m==lc[s]) continue;
mc[j]=s;
j++; }
return mc; }
}
```

1.3.   *Cutset matrix*

Cutset is the minimal edge set that makes a connected graph unconnected. Cutset matrix describes vertex-edge relationship of graph.

An undirected and connected graph X with v vertices and e edges, has a cutset matrix $Q = (q_{ij})_{k \times e}$, where k is the total number of cutsets. If the edge e_j is in the cutset k_i, then $q_{ij} = 1$ and if the edge e_j is not in the cutset k_i then $q_{ij} = 0$; where $i = 1, 2, \ldots, k$; $j = 1, 2, \ldots, e$.

For example, there are seven cutsets in the graph of Fig. 2, and the cutset matrix is:

$$Q = (q_{ij})_{7 \times 6} = \begin{bmatrix} 1 & 0 & 1 & 1 & 0 & 0 \\ 1 & 1 & 0 & 0 & 0 & 1 \\ 0 & 1 & 1 & 0 & 1 & 0 \\ 0 & 0 & 0 & 1 & 1 & 1 \\ 0 & 1 & 1 & 1 & 0 & 1 \\ 1 & 0 & 1 & 0 & 1 & 1 \\ 1 & 1 & 0 & 1 & 1 & 0 \end{bmatrix}.$$

An undirected and connected graph X with v vertices and e edges has its fundamental cutset matrix as $Q = (q_{ij})_{(v-1) \times e}$, which responds to a tree and each row responds to a fundamental cutset and each column responds to an edge. If the edge e_j is in the fundamental cutset k_i, then $q_{ij} = 1$ and if the edge e_j is not in the fundamental cutset k_i then $q_{ij} = 0$; $i = 1, 2, \ldots, k$; $j = 1, 2, \ldots, e$. The rank of fundamental cutset matrix Q; is $v - 1$.

If Q and C are cutset matrix and circuit matrix of an undirected and connected graph respectively, and the edges represented by columns of two matrices are the same, then $QC^T = 0$, $CQ^T = 0$.

For an undirected and connected graph X with v vertices and e edges, an algorithm to calculate the cutset matrix Q of X is (Chan *et al.*, 1982):

(1) Choose a tree T from a graph X, and write out the fundamental cutset matrix Q_a of tree T.
(2) Make all possible ring sum operations in Q_a, and construct a new matrix Q_b by adding original rows of Q_a, together with the new rows generated from ring sum operations.
(3) Eliminate redundant rows (i.e., the rows of the circuits with disjoint edges) in Q_b.
(4) Construct the cutset matrix Q from the left rows of Q_b.

The algorithm for cutset matrix Q is the same as the algorithm for circuit matrix C.

1.4. *Calculation of fundamental matrix*

To analyze incidence matrix, circuit matrix and cutset matrix, we need to calculate the fundamental matrix. The following are Java codes, elemTrans, for calculating fundamental matrix and its rank through elementary transformation:

```
//Calculate fundamental matrix.
/*n: number of vertice; qa[n][m]: known matrix. q[n][m]:
fundamental matrix. */
  public class elemTrans {
  public static void main(String[] args){
  int i,j,v,e;
```

```
if (args.length!=2)
System.out.println("You must input the name of table in the
database. For example, you may type the following in the command
window: java elemTrans 98 elemtrans, where elemtrans is the name
of table, 98 means the number of edges.");
e=Integer.valueOf(args[0]).intValue();
String tablename=args[1];
readDatabase                                        readdata=new
readDatabase("dataBase",tablename,e);
v=readdata.m;
int a[][]=new int[v+1][e+1];
for(i=1;i<=v;i++)
for(j=1;j<=e;j++)
a[i][j]=(Integer.valueOf(readdata.data[i][j])).intValue();
elemTrans(v,e,a); }
public static int[][] elemTrans(int n, int m, int qa[][]) {
int i,j,k,l,s,t;
int q[][]=new int[n+1][m+1];
int e[]=new int[n+1];
int ee[]=new int[n+1];
int c[]=new int[m+1];
s=0;
t=m;
for(i=1;i<=n;i++) {
e[i]=0;
ee[i]=0; }
for(j=1;j<=m;j++)
for(i=1;i<=n;i++) q[i][j]=0;
loop: for(j=1;j<=m;j++) {
for(i=1;i<=n;i++) {
if ((ee[i]!=0) | (qa[i][j]==0)) continue;
s++;
c[s]=j;
e[s]=i;
ee[i]=s;
for(k=1;k<=n;k++) {
if ((k==i) | (qa[k][j]==0)) continue;
for(l=1;l<=m;l++) {
qa[k][l]+=qa[i][l];
if (qa[k][l]==2) qa[k][l]=0; }
continue loop; } }
c[t]=j;
t--; }
for(j=1;j<=m;j++)
for(i=1;i<=s;i++) {
k=e[i];
```

```
l=c[j];
q[i][j]=qa[k][l];}
System.out.println("Fundamental matrix:");
for(i=1;i<=s;i++) {
for(j=1;j<=m;j++)
System.out.print(q[i][j]+" ");
System.out.println(); }
System.out.println("Rank of fundamental matrix: "+s);
return q; }
}
```

2. Directed Graph

2.1. *Incidence matrix*

A directed graph X with v vertices and e edges, has an incidence matrix $A = (a_{ij})_{v \times e}$, where $a_{ij} = 1$, if $(v_i, v_k) = e_j$; $a_{ij} = -1$, if $(v_k, v_i) = e_j$, and $a_{ij} = 0$, if $(v_i, v_k) \neq e_j$, $(v_k, v_i) \neq e_j$; $i = 1, 2, \ldots, v$; $j = 1, 2, \ldots, e$.

For example, the incidence matrix A of the graph in Fig. 2, is

$$A = (a_{ij})_{4 \times 6} = \begin{bmatrix} -1 & 0 & 1 & 1 & 0 & 0 \\ 1 & 1 & 0 & 0 & 0 & 1 \\ 0 & -1 & -1 & 0 & -1 & 0 \\ 0 & 0 & 0 & -1 & 1 & -1 \end{bmatrix}.$$

The rank of incidence matrix of a directed graph with v vertices is $v - 1$. For a directed graph with v vertices and p maximal connected subgraphs, the rank of its incidence matrix is $v - p$.

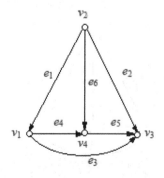

Figure 2.   A directed graph.

2.2. *Adjacency matrix*

Adjacency matrix describes vertex-vertex relationship of graph. It is mostly used in directed graphs, but is also suitable to undirected graphs. Adjacency matrix has many applications. It may represent some properties of graph, such as the loop of a vertex (whether the corresponding diagonal element is 1), whether the edges are pairly occurred (whether the matrix is symmetric), and the outdegree and indegree of a vertex (row sum and column sum of matrix), etc. In addition, the properties of a graph can be approached by matrix operations.

An undirected graph X with v vertices and e edges, has adjacency matrix $D = (d_{ij})_{v \times v}$. If v_i and v_j are adjacent, then $d_{ij} = 1$ and $d_{ij} = 0$, if v_i and v_j are not adjacent; where $i, j = 1, 2, \ldots, v$.

For example, the adjacency matrix of the graph in Fig. 3 is

$$D = (d_{ij})_{4 \times 4} = \begin{bmatrix} 0 & 1 & 1 & 1 \\ 1 & 0 & 1 & 1 \\ 1 & 1 & 0 & 1 \\ 1 & 1 & 1 & 0 \end{bmatrix}.$$

The number of 1's in a row or column of adjacency matrix is the degree of the corresponding vertex.

If an adjacency matrix is a symmetric matrix with all diagonal elements as 0, then there is no loop in the graph.

In the elementary transformation of adjacency matrix, permuting a row means that the corresponding column must be also permuted. Suppose there

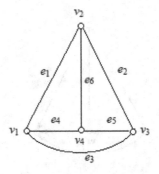

Figure 3.   An undirected connected graph X.

exists a permutation matrix P, such that

$$D_2 = P^{-1} D_1 P,$$

then the graphs corresponding to D_2 and D_1 are isomorphic.

If there is an adjacency matrix D

$$D = \begin{bmatrix} D_{11} & 0 \\ 0 & D_{22} \end{bmatrix}$$

then the graph is an unconnected graph, and it has at least two maximal connected subgraphs.

Adjacency matrix can be generalized to the multigraphs with parallel edges and weighted graphs. For example, the value of element d_{ij} can be the type of the edge from v_i to v_j, or can be the weight w_{ij} of the edge; and if there is no edge from v_i to v_j, then let $d_{ij} = 0$.

2.3. *Circuit matrix*

A directed and connected graph X with v vertices, e edges and c circuits, has a circuit matrix $C = (c_{ij})_{c \times e}$. If the edge e_j is in the circuit c_i, and in the same direction of the circuit, then $c_{ij} = 1$; $c_{ij} = -1$, if the edge e_j is in the circuit c_i, and in the opposite direction of the circuit, and $c_{ij} = 0$, if the edge e_j is not present in the circuit c_i; $i = 1, 2, \ldots, c$; $j = 1, 2, \ldots, e$.

Under some rule for circuit direction, for example, the circuit matrix of the graph in Fig. 3 is

$$C = (c_{ij})_{7 \times 6} = \begin{bmatrix} 1 & 0 & 0 & 1 & 0 & -1 \\ 0 & -1 & 0 & 0 & 1 & 1 \\ 0 & 0 & -1 & 1 & 1 & 0 \\ 1 & -1 & 1 & 0 & 0 & 0 \\ 1 & -1 & 0 & 1 & 1 & 0 \\ 1 & 0 & 1 & 0 & -1 & -1 \\ 0 & 1 & -1 & 1 & 0 & -1 \end{bmatrix}.$$

The rank of both circuit matrix and fundamental circuit matrix of a directed and connected graph X with v vertices and e edges is $e - v + 1$.

If A and C are the incidence matrix and circuit matrix of a directed and connected graph X respectively, and the edges represented by columns of two matrices are the same, then $AC^T = 0$, $CA^T = 0$.

2.4.   *Cutset matrix*

A directed and connected graph X with v vertices and e edges, has a cutset matrix $Q = (q_{ij})_{k \times e}$, where k is the total number of cutsets. If the edge e_j is in the cutset k_i, and in the same direction of the cutset, then $q_{ij} = 1$; $q_{ij} = -1$, if the edge e_j is in the cutset k_i, and in the opposite direction of the cutset, and $q_{ij} = 0$, if the edge e_j is not present in the cutset k_i; where $i = 1, 2, \ldots, k$; $j = 1, 2, \ldots, e$.

If Q and C are cutset matrix and circuit matrix of a directed and connected graph X respectively, and the edges represented by columns of two matrices are the same, then $QC^{\mathrm{T}} = 0$, $CQ^{\mathrm{T}} = 0$.

2.5.   *Walk matrix and reachability matrix*

Let the adjacency matrix of a graph X with n vertices is D, $\vee$ and $\wedge$ represent the following defined matrix operations. If $D = (d_{ij})$, $C = (c_{ij})$, then

$$D \vee C = (a_{ij}) \quad a_{ij} = d_{ij} \vee c_{ij}$$
$$D \wedge C = (b_{ij}) \quad b_{ij} = \vee_{k=1}^{n} d_{ik} \wedge c_{kj}$$

Let $D^{(m)}$ be $D \wedge D \wedge \cdots \wedge D$. Consider the matrix

$$F = D \vee D^{(2)} \vee \cdots \vee D^{(n)}$$

$f_{ij} = 1$, if and only if there exists a path (walk) from v_i to v_j in X. The walk matrix of graph X is denoted by F.

Let $P = I \vee F$ where $I_{n \times n}$ is the unit matrix, P is thus called reachability matrix of graph X.

For example, calculate the walk matrix F and reachability matrix P of the graph in Fig. 4.

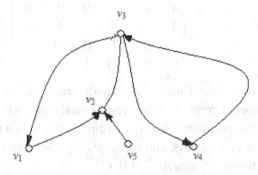

Figure 4.   A directed graph.

Suppose the adjacency matrix of the graph is D, then

$$D = \begin{bmatrix} 0 & 1 & 0 & 0 & 0 \\ 0 & 0 & 1 & 0 & 0 \\ 1 & 0 & 0 & 1 & 0 \\ 0 & 0 & 1 & 0 & 0 \\ 0 & 1 & 0 & 0 & 0 \end{bmatrix}$$

$$D^{(2)} = \begin{bmatrix} 0 & 0 & 1 & 0 & 0 \\ 1 & 0 & 0 & 1 & 0 \\ 0 & 1 & 1 & 0 & 0 \\ 1 & 0 & 0 & 1 & 0 \\ 1 & 0 & 1 & 0 & 0 \end{bmatrix}.$$

Thus the walk matrix F and reachability matrix P are as follows

$$F = D \vee D^{(2)} \vee D^{(3)} \vee D^{(4)} \vee D^{(5)} = \begin{bmatrix} 1 & 1 & 1 & 1 & 0 \\ 1 & 1 & 1 & 1 & 0 \\ 1 & 1 & 1 & 1 & 0 \\ 1 & 1 & 1 & 1 & 0 \\ 1 & 1 & 1 & 1 & 0 \end{bmatrix}$$

$$P = I \vee F = \begin{bmatrix} 1 & 1 & 1 & 1 & 0 \\ 1 & 1 & 1 & 1 & 0 \\ 1 & 1 & 1 & 1 & 0 \\ 1 & 1 & 1 & 1 & 0 \\ 1 & 1 & 1 & 1 & 1 \end{bmatrix}.$$

Reachability matrix P shows that the graph in Fig. 4 is a unidirected and connected graph.

3.   Computer Storage of Graph

A matrix containing a large number of 0's should be stored in the way used by sparse matrix storage.

Binary matrix may be used to store undirected graph. In addition, the adjacency matrix of an undirected graph is a symmetric matrix, so just the upper triangular matrix needs to be stored.

In addition to matrix, a graph can also be stored as pairs of vertices. For example, the graph in Fig. 2 can be stored in pairs of vertices: (1,2), (2,3), (3,1), (1,4), (4,3), (2,4).

The principle for storing graph with two linear array is to define two arrays with the same dimension, A_1 and A_2; an element of A_1 stores a vertex of an edge and the corresponding element of A_2 stores another vertex of the edge. If an edge weight needs to be stored, the third array should be used. For example, the representation of two linear array of the graph in Fig. 2 is

$$A_1 = (1, 2, 3, 1, 4, 2)$$
$$A_2 = (2, 3, 1, 4, 3, 4).$$

Successor listing is often used to the graphs with many vertices and fewer edges. Define two arrays with the same dimension, A_1 and A_2; A_1 stores vertices and A_2 stores between-vertex adjacency relation. For example, $A_1(i) = v_i$, $A_2(i + 1) = v_{i+1}$, then the elements of A_2 starting from v_ith element, are vertices adjacent to v_i. Thus the elements from $A_2(v_i)$ to $A_2(v_{i+1})$, are the vertices associated with the vertex i. The successor listing of the graph in Fig. 2 is

$$A_1 = (1, 4, 6, 9, 12)$$
$$A_2 = (2, 3, 4, 1, 3, 4, 1, 2, 4, 1, 2, 3).$$

Adjacency vertice listing is usually used in DFS algorithm. It is applicable to the situation with small ratio of edges *vs.* vertices. Two arrays are used in this representation. The one-dimensional array R marks the degree of every vertex, and two-dimensional array $P = (p_{ij})_{v \times d}$ marks the vertices adjacent to each vertex, where v is the number of vertices, p_{ij} is the labeled number of jth vertex adjacent to the vertex i in which the vertices adjacent to vertex i can be ordered arbitrarily. The one-dimensional array R and two-dimensional array P of the graph in Fig. 3 is

$$(3, 3, 3, 3)$$

$$\begin{bmatrix} 2 & 3 & 4 \\ 1 & 3 & 4 \\ 1 & 2 & 4 \\ 1 & 2 & 3 \end{bmatrix}$$

The Java codes, adjListAdjMat, for storing adjacency matrix with adjacency vertice listing are

```
/*Transform adjacency matrix to the matrix stored in Adjacency
Vertice Listing.*/
  /*v: number of vertice; p[1—v][0]: 1-dim degree vector; p[1—v]
[1—dd]: 2-dim matrix stored in Adjacency Vertice Listing; dd:
the maximum degree of vertex. */
  public class adjListAdjMat {
  public int dd,p[][];
  public void dataTrans(int d[][]) {
  int i,j,k,v,num;
  v=d.length—1;
  int nu[]=new int[v+1];
  for(i=1;i<=v;i++)
  for(j=1;j<=v;j++)
  if (d[i][j]!=0) d[i][j]=1;
  dd=0;
  for(i=1;i<=v;i++) {
  num=0;
  for(j=1;j<=v;j++) num+=d[i][j];
  nu[i]=num;
  if (num>dd) dd=num; }
  p=new int[v+1][dd+1];
  for(i=1;i<=v;i++) p[i][0]=nu[i];
  for(i=1;i<=v;i++) {
  j=0;
  for(k=1;k<=v;k++)
  if (d[i][k]==1) {
  j++;
  p[i][j]=k; }
  for(k=j+1;k<=dd;k++) p[i][k]=0; } }
  }
```

4. Ecological Applications

4.1. *Simple ecological relationship and incidence matrix*

Simple ecological relationship is always represented by square matrix. In species interactions or food web studies, for example, species interactions or food web constitute a directed graph. Suppose there are m species and the raw data are (r_{ij}), $i, j = 1, 2, \ldots, m$, where r_{ij} is the interaction type of species i to species j (e.g., 1 denotes predation or parasitism, -1

denotes predated or parasitized, 0 denotes neutral interaction or null inter-action). We need to transform the matrix $R = (r_{ij})_{m \times m}$ to incidence matrix $A = (a_{ij})_{m \times e}$, where e is the total number of non-zero elements in upper triangular matrix (i.e., the number of edges). For example, consider the following two matrices:

$$R = (r_{ij})_{6\times 6} = \begin{bmatrix} 0 & 0 & -1 & 1 & 0 & -1 \\ 0 & 0 & 0 & 0 & -1 & 1 \\ 1 & 0 & 0 & 1 & 0 & -1 \\ -1 & 0 & -1 & 0 & 1 & 0 \\ 0 & 1 & 0 & -1 & 0 & 0 \\ 1 & -1 & 1 & 0 & 0 & 0 \end{bmatrix}$$

$$A = (a_{ij})_{6\times 8} = \begin{bmatrix} -1 & 1 & -1 & 0 & 0 & 0 & 0 & 0 \\ 0 & 0 & 0 & -1 & 1 & 0 & 0 & 0 \\ 1 & 0 & 0 & 0 & 0 & 1 & -1 & 0 \\ 0 & -1 & 0 & 0 & 0 & -1 & 0 & 1 \\ 0 & 0 & 0 & 1 & 0 & 0 & 0 & -1 \\ 0 & 0 & 1 & 0 & -1 & 0 & 1 & 0 \end{bmatrix}.$$

If species interactions or food web constitutes an undirected graph, the matrix transformation is similar to the above.

The Java codes, netMatIncMat, for transformating species interactions to incidence matrix are

```
/*Transform eco-net to incidence matrix. */
/*v: number of vertice; e: number of edges; r[1-v][1-v]: eco-net
matrix; a[1-v][1-e]: incidence matrix; The type is 0 if the graph is
a non-oriented graph and is 1 if it is an oriented graph. */
public class netMatIncMat {
public int v,e,a[][];
public void dataTrans(int type, int r[][]) {
int i,j,k;
v=r.length-1;
e=0;
for(i=1;i<=v-1;i++)
for(j=i+1;j<=v;j++) if (r[i][j]!=0) e++;
a=new int[v+1][e+1];
for(i=1;i<=v;i++)
for(j=1;j<=e;j++) a[i][j]=0;
k=0;
for(i=1;i<=v-1;i++)
```

```
for(j=i+1;j<=v;j++)
if (r[i][j]!=0) {
k++;
a[i][k]=r[i][j];
if (type==1) a[j][k]=-r[i][j];
else a[j][k]=r[i][j]; } }
}
```

The above species interaction relationship has not included loops and parallel edges. For the relationship containing loops and parallel edges, if it is stored with two linear array, then the Java codes, netVecIncMat, for transforming to incidence matrix are

```
/*Transform eco-net (represented in two linear arrays) to
incidence matrix. The type is 0 if the graph is a non-oriented
graph and is 1 if it is an oriented graph. e: number of edges; v:
number of vertice. All vertice in the graph must be labeled
sequentially (from 1 to v). a1[1-e]: the set for from-vertex;
a2[1-e]: the set for to-vertex; aa[1-e]: 1 if it is from vertex
a1[1-e] to vertex a2[1-e], -1 if from vertex a2[1-e] to vertex
a1[1-e]; m: number of vertice. a[1-v][1-e]: incidence matrix*/
public class netVecIncMat {
public int v,e,a[][];
public void dataTrans(int type, int a1[], int a2[], int aa[]) {
int i,j;
e=a1.length-1;
v=0;
for(i=1;i<=e;i++) {
if (a1[i]>v) v=a1[i];
if (a2[i]>v) v=a2[i]; }
a=new int[v+1][e+1];
for(i=1;i<=v;i++)
for(j=1;j<=e;j++) a[i][j]=0;
for(i=1;i<=e;i++) {
a[a1[i]][i]=aa[i];
if (type==1) a[a2[i]][i]=-aa[i];
else a[a2[i]][i]=aa[i]; } }
}
```

4.2. *Ecological relationship and adjacency matrix*

In a simple ecological relationship matrix R, by taking absolute values of all elements, it becomes the adjacency matrix D, i.e., $D = R = (r_{ij})$, if R

is a matrix for undirected graph and $D = R = (|r_{ij}|)$, if R is a matrix for directed graph.

Therefore in a sense the above ecological relationship matrices are adjacency matrices.

The adjacency matrix above can only represent simple ecological relationship. To represent complex ecological relationship, it is necessary to redefine adjacency matrix.

In this book, we suppose that the adjacency matrix is D, which represents the type and property of between-vertex edges. If $d_{ij} = d_{ji} = 0$, then there is no edge from v_i to v_j; if $d_{ij} = -d_{ji}$, and $|d_{ij}| = 1$, then there is only a directed edge from v_i to v_j; if $d_{ij} = d_{ji} = 1$, then there is only an undirected edge from v_i to v_j; if $d_{ij} = d_{ji} = 2$, then there are two parallel edges from v_i to v_j; if $d_{ii} = 3$, then v_i has a loop; if $d_{ii} = 4$, then v_i is an isolated vertex; if $d_{ii} = 5$, then v_i is an isolated vertex and it has a loop.

For a weighted graph, a weight matrix, D_w, of the adjacency matrix D, should also be added in which the elements are weights of edges and weights take positive or negative values as the direction of edges.

For the weighted graph in Fig. 5, the adjacency matrix D is

$$\begin{bmatrix} 0 & 1 & -1 & 1 & 0 & 0 \\ -1 & 3 & 1 & 1 & 0 & 0 \\ 1 & -1 & 0 & 2 & 0 & 0 \\ -1 & -1 & 2 & 0 & 0 & 0 \\ 0 & 0 & 0 & 0 & 5 & 0 \\ 0 & 0 & 0 & 0 & 0 & 4 \end{bmatrix}$$

and the weight matrix D_w is

$$\begin{bmatrix} 0 & 8 & -9 & 5 & 0 & 0 \\ -8 & 5 & 4 & 7 & 0 & 0 \\ 9 & -4 & 0 & 2 & 0 & 0 \\ -5 & -7 & 6 & 0 & 0 & 0 \\ 0 & 0 & 0 & 0 & 2 & 0 \\ 0 & 0 & 0 & 0 & 0 & 0 \end{bmatrix}.$$

For the situation that there are m edges of the same direction between two vertices, we can replace m edges of the same direction with an edge in the

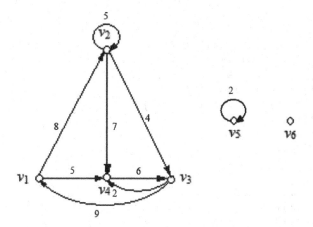

Figure 5.   A directed graph.

same direction, and the gain of the edge is the sum of weights (or minimal weight, or maximal weight, etc.) of the m edges of the same direction. For the situation that a vertex has several loops, the method is the same.

As a result, there are at most two edges between two vertices and a vertex has at most a loop.

The Java codes, twoArrAdjMat, for transforming adjacency matrix to two linear array representation are

```
/*Transform adjacency matrix to the data in Two Arrays Listing.*/
/*v: number of vertice; e: number of edges; d1[1-e], d2[1-e]:
start and end vertice; d[1-e]: values of edges; d[1-v][1-v]:
adjacency matrix to reflect the feature of edges, e.g., dij=dji=0
means no edge between vertice i and j; dij=-dji, and |dij|=1, means
there is an edge between vertice i and j; dij=dji=2, means there are
paralle edges between vertice i and j; dii=3 means there is a self-
loop for vertex i; dii=4 means isolated vertex; dii=5 means isolated
vertex i with self-loop.ddw[1-v][-v]: weight matrix; weights are
signed according to the signs of edges. */
public class twoArrAdjMat {
public int e,v,d1[],d2[],dd[];
public double dw[];
public void dataTrans(int d[][]) {
int i,j,n;
int i1,i2,i3,i4,i5;
i1=i2=i3=i4=i5=0;
v=d.length-1;
int nu[]=new int[v+1];
```

```
for(i=1;i<=v;i++)
for(j=1;j<=v;j++) {
if ((Math.abs(d[i][j])==1) & (i<j)) i1++;
if ((d[i][j]==2) & (i<j)) i2++;
if ((d[i][j]==3) & (i==j)) i3++;
if ((d[i][j]==4) & (i==j)) i4++;
if ((d[i][j]==5) & (i==j)) i5++; }
e=i1+i2+i3+i4+i5;
d1=new int[e+1];
d2=new int[e+1];
dd=new int[e+1];
n=0;
for(i=1;i<=v;i++)
for(j=1;j<=v;j++) {
if ((Math.abs(d[i][j])==1) & (i<j)) {
n++;
d1[n]=i;
d2[n]=j;
dd[n]=d[i][j]; }
if ((d[i][j]==2) & (i<j)) {
n++;
d1[n]=i;
d2[n]=j;
dd[n]=2; }
if ((d[i][j]==3) & (i==j)) {
n++;
d1[n]=i;
d2[n]=j;
dd[n]=3; }
if ((d[i][j]==4) & (i==j)) {
n++;
d1[n]=i;
d2[n]=j;
dd[n]=4; }
if ((d[i][j]==5) & (i==j)) {
n++;
d1[n]=i;
d2[n]=j;
dd[n]=5; } } }
public void dataTrans(int d[][], double ddw[][]) {
int i,j,n;
dataTrans(d);
dw=new double[e+1];
n=1;
for(i=1;i<=v;i++)
for(j=1;j<=v;j++) {
```

```
if ((Math.abs(d[i][j])==1) & (i<j)) {
dw[n]=ddw[i][j];
n++; }
if ((d[i][j]>=2) && (d[i][j]<=5)) {
dw[n]=ddw[i][j];
n++; } } }
}
```

Transforming the above adjacency matrix D to two linear array representation, the results are

$$D_1 = (1, 1, 1, 2, 2, 2, 3, 4, 5, 6);$$
$$D_2 = (2, 3, 4, 2, 3, 4, 4, 3, 5, 6);$$
$$Dd = (1, -1, 1, 3, 1, 1, 2, 2, 5, 4);$$
$$Dw = (8, -9, 5, 5, 4, 7, 2, 6, 2, 0).$$

The Java codes, adjMatTwoArr, for transforming two linear array representation to adjacency matrix and weight matrix are

```
/*Transform the data in TwoArraysListing to adjacency matrix.*/
/*v: number of vertice; e: number of edges; d1[1−e], d2[1−e]:
start and end vertice; dd[1−e]: values of edges, 1 if it is a forward
edge; −1 is for backward edge; 4 if s1[i]=s2[i], i.e., the vertex
is an isolated vertex with no self-loop; 2: parallel edges; 3: self-
loop for non-isolated vertex; 5: an isolated vertex with self-
loop. d[1−v][1−v]: adjacency matrix to reflect the feature of
edges, e.g., dij=dji=0 means no edge between vertice i and j;
dij=−dji, and |dij|=1, means there is an edge between vertice
i and j; dij=dji=2, means there are paralle edges between vertice
i and j; dii=3 means there is a self-loop for vertex i; dii=4 means
isolated vertex; dii=5 means isolated vertex i with self-loop. */
public class adjMatTwoArr {
public int v,d[][];
public void dataTrans(int d1[], int d2[], int dd[]) {
int i,j,k,e;
e=d1.length−1;
v=0;
for(i=1;i<=e;i++) {
if (d1[i]>v) v=d1[i];
if (d2[i]>v) v=d2[i]; }
d=new int[v+1][v+1];
for(i=1;i<=v;i++)
for(j=1;j<=v;j++)
for(k=1;k<=e;k++)
```

```
if ((d1[k]==i) & (d2[k]==j)) {
d[i][j]=dd[k];
if (Math.abs(d[i][j])==1) d[j][i]=-d[i][j];
if (d[i][j]==2) d[j][i]=2;
break; } }
}
```

4.3.   *Matrix transformation algorithm*

The following algorithm loads all the above transformation algorithms and generates needed data:

```
public class dataTrans {
public static void main(String[] args){
int i,j,m,n,e,v,type;
if (args.length!=3)
System.out.println("You must input the number of columns of table
in the database, type of data transformation, and the name of table
in the database.
For example, you may type the following in the command window:
java dataTrans 3 5 datatrans5, where datatrans5 is the name of
table, 3 means the number of data columns in the table, 5 means
transforming the data in Two Arrays Listing to adjacency matrix.");
System.out.println("Type of data transformation: transform
adjacency matrix to the data in Two Arrays Listing (1); Transform
adjacency matrix to the matrix stored in Adjacency Vertice Listing
(2); Transform eco-net to incidence matrix (3); Transform eco-net
(represented in two linear arrays) to incidence matrix (4);
Transform the data in Two Arrays Listing to adjacency matrix (5).
\n");
n=Integer.valueOf(args[0]).intValue();
type=Integer.valueOf(args[1]).intValue();
String tablename=args[2];
readDatabase                                  readdata=new
readDatabase("dataBase",tablename,n);
m=readdata.m;
double a[][]=new double[m+1][n+1];
int b[][]=new int[m+1][n+1];
for (i=1;i<=m;i++)
for (j=1;j<=n;j++) {
a[i][j]=(Double.valueOf(readdata.data[i][j])).doubleValue();
b[i][j]=(int)a[i][j]; }
if (type==1) {
twoArrAdjMat arr=new twoArrAdjMat();
arr.dataTrans(b);
```

```
e=arr.e;
for(i=1;i<=e;i++)
System.out.print(arr.d1[i]+" "+arr.d2[i]+"
"+arr.dd[i]+"\n"); }
if (type==2) {
adjListAdjMat arr=new adjListAdjMat();
arr.dataTrans(b);
e=arr.dd;
System.out.println("Degree of every vertice:");
for(i=1;i<=n;i++) System.out.print(arr.p[i][0]+" ");
System.out.println();
System.out.println("Matrix:");
for(i=1;i<=n;i++) {
for(j=1;j<=e;j++)
System.out.print(arr.p[i][j]+" ");
System.out.println(); } }
if (type==3) {
int typ=0;
for(i=1;i<=m;i++)
for(j=1;j<=n;j++)
if (b[i][j]<0) {
typ=1;
break; }
netMatIncMat arr=new netMatIncMat();
arr.dataTrans(typ,b);
e=arr.e;
for(i=1;i<=n;i++) {
for(j=1;j<=e;j++)
System.out.print(arr.a[i][j]+" ");
System.out.println(); } }
if (type==4) {
int typ=0;
int a1[]=new int[m+1];
int a2[]=new int[m+1];
int aa[]=new int[m+1];
for(i=1;i<=m;i++) {
a1[i]=b[i][1];
a2[i]=b[i][2];
aa[i]=b[i][3];
if (aa[i]<0) typ=1; }
netVecIncMat arr=new netVecIncMat();
arr.dataTrans(typ,a1,a2,aa);
v=arr.v;
for(i=1;i<=v;i++) {
for(j=1;j<=m;j++)
System.out.print(arr.a[i][j]+" ");
```

```
System.out.println(); } }
if (type==5) {
int d1[]=new int[m+1];
int d2[]=new int[m+1];
int dd[]=new int[m+1];
for(i=1;i<=m;i++) {
d1[i]=b[i][1];
d2[i]=b[i][2];
dd[i]=b[i][3]; }
adjMatTwoArr arr=new adjMatTwoArr();
arr.dataTrans(d1,d2,dd);
v=arr.v;
for(i=1;i<=v;i++) {
for(j=1;j<=v;j++)
System.out.print(arr.d[i][j]+" ");
System.out.println(); } }
} }
```

4.4. *Read ecological relationship matrix from database*

Ecological relationship data are always stored in database. JDBC supports various databases, such as SQL, Oracle, FoxPro, Excel, Access, etc. The following Java codes, readDatabase, are designed to read ecological relationship matrix from database:

```
/*Read matrix data from JDBC databases (Excel, Access, Oracle,
SQL, etc.). database: name of database; table: name of table; m:
number of rows of matrix data; n: number of columns of matrix data. */
/*data[1-m][1-n]: returned matrix data. */
import java.sql.*;
import java.net.URL;
public class readDatabase {
private Connection con;
private Statement st;
private String strr;
public double data[][];
public int m;

public readDatabase(String database,String table,int n) {
int k;
try {
String driver_name="sun.jdbc.odbc.JdbcOdbcDriver";
String database_name="jdbc:odbc:"+database;
Class.forName(driver_name);
con=DriverManager.getConnection(database_name);
```

```
st=con.createStatement();
String stt="SELECT * FROM "+table;
ResultSet rs=st.executeQuery(stt);
m=0;
while (rs.next()) {
m++; }
rs.close();
data=new double[m+1][n+1];
ResultSet rss=st.executeQuery(stt);
k=0;
while (rss.next()) {
k++;
for(int i=1;i<=n;i++)
data[k][i]=(Double.valueOf(rss.getString(i))).doubleValue(); }
rss.close(); }
catch(ClassNotFoundException e)
{System.out.println("Class not found!"); }
catch(Exception e)
{System.out.println(e.getMessage()); } }
}
```

❧ CHAPTER 4 ❧

Trees and Planar Graphs

1. Tree

1.1. *Tree*

The graph without any circuit is called an acyclic graph. The connected acyclic graph is called a tree. Decision tree, ranking numbers, operation sequence of expression, and so on, are all examples of the tree.

A tree is called the spanning tree of a graph X, if the tree contains all vertices of X. A connected graph must contain a spanning tree.

The complementary subgraph of a tree is called a cotree of the tree. The branches of a cotree are called the chords. If T is the tree of a graph X, then the subgraph obtained by removing the edges of T from X is the cotree of T.

By adding a chord to a tree of the connected graph X, we form a closed path that has only one chord and the others are tree branches. The single-chord circuit generated in this way is called a fundamental circuit. The number of fundamental circuits is equivalent to the number of chords.

Corresponding to an edge that belongs to a cutset of a tree in connected graph X, if there is only one tree branch and the others are all chords, the single- branch cutset is called a fundamental cutset. The number of fundamental cutsets is equivalent to the number of tree branches.

A cut-edge of a graph X refers to the edge e such that $\omega(X-e) > \omega(X)$. A connected graph is a tree if and only if all of its edges are cut-edges.

Theorem 1. *A graph is a tree, if and only if there is one and only one path between any two edges.*

Theorem 2. *There exists at least one tree in a connected graph.*

Theorem 3. *The tree of a connected graph with v vertices must contain v − 1 edges.*

Theorem 4. *Given a connected graph X with v vertices, $X_i \subset X$ has v − 1 edges and if there is not even a single circuit in it, then X_i is a tree of X.*

Theorem 5. *The tree of a connected graph X with v vertices and e edges must contain and only contains v − 1 tree branches and e − v + 1 chords.*

Theorem 6. *If e is a linkage of a graph X, then the number of spanning trees of X is $\tau(X) = \tau(X - e) + \tau(X \bullet e)$, where $X \bullet e$ refers to the graph that eliminates e from X and makes two endpoints of e the same point.*

Theorem 7. *Let the incidence matrix of a connected graph X be B, the number of spanning trees of X is $det(BB^T)$.*

1.2.   *Ingoing tree, outgoing tree*

Ingoing tree and outgoing tree are two types of special trees (Fig. 1).

For a directed tree T, if there exists a vertex v_0, such that the other vertex v of the tree always has a relation $v_0 < v$, then the tree is called an outgoing tree, where $v_0 < v$ means that there exists a path from v_0 to v. The characteristic of an outgoing tree is that besides the starting point v_0, every vertex has one and only one edge to reach it, i.e., indegree of the vertex is one.

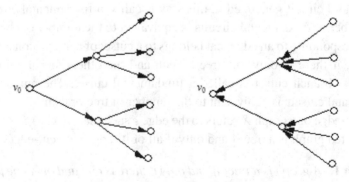

Figure 1.   Outgoing tree and ingoing tree.

Figure 2.   A binary tree.

For a directed tree T, if there exists a vertex v_0, such that the other vertex v of the tree always has a relation $v < v_0$, then the tree is called an ingoing tree with v_0 as the endpoint, where $v_0 < v$ means that there exists a path from v_0 to v. The characteristic of ingoing tree is that besides the endpoint v_0, every vertex has one and only one edge to leave it, i.e., outdegree of the vertex is one.

1.3.   *Binary tree*

Binary tree is an outgoing tree (Fig. 2). The vertex with zero indegree in an outgoing tree is called a tree root and the indegree of all other vertices is one. The characteristic of a binary tree is that besides leaf vertices with zero outdegree, the outdegree of all other vertices is not greater than two. The vertices with non-zero outdegree are called branching vertices.

2.   Planar Graph

Planar graph is a short name for planary graph. If a graph that appears to be a nonplanar graph but can be redrawn such that any two edges intersect only at their endpoints, then it must be a planar graph.

2.1.   *Planar graph, Nonplanar graph*

A graph is a planar graph if any two edges intersects only at their endpoints, otherwise it is a nonplanar graph. All planar domains determined by planar graph are called faces. Bounded domain is called an internal face and unbounded domain is called an external face. There is only one external face.

Suppose graph X is a planar graph without loops and parallel edges, v_i and v_j are any two disjoint vertices. X is called the maximal planar graph if it is impossible to add an edge between v_i and v_j but will not break the planarity of graph X. Each face of maximal planar graph is surrounded by three edges. A graph X may become maximal planar graph by adding edges one by one to X.

According to Chapter 2, if there exists a one-to-one correspondence map, $f : R^2 \rightarrow \mathcal{K}$, between the vertices of R^2 and graph $\mathcal{K}$ (one-dimensional abstract complex), such that $(a^{i0}, a^{i1}, \ldots, a^{ir})$ is a simplex of R^2 if and only if $(f(a^{i0}), f(a^{i1}), \ldots, f(a^{ir}))$ is the abstract simplex of $\mathcal{K}$, then $\mathcal{K}$ is a planar graph, and R^2 is the geometric realization of $\mathcal{K}$.

If X is a planar graph, all of its subgraphs are also planar graphs.

Kuratowski Theorem. *A graph X is a planar graph if and only if X does not contain any of the two Kuratowski graphs or its homeomorphic graph for the subgraph of X.*

Kuratowski graph is also called fundamental nonplanar graph. It has two forms, as indicated in Fig. 3.

Theorem 8 (Euler Formula). *If graph X is a connected and planary graph with v vertices, e edges, and f faces, then $v - e + f = 2$.*

If graph X is a connected and planar graph with v vertices ($v \geq 3$), e edges and without loops and parallel edges, then $e \leq 3v - 6$.

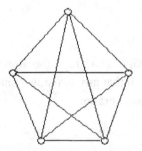 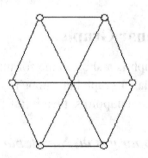

Figure 3.   Kuratowski graphs.

Theorem 9. *If graph X is a planar graph without loops and parallel edges, then there is at least one vertex whose degree is less than six.*

Theorem 10. *The number of vertices v, edges e and faces f of maximal planar graph satisfies the equalities: $e = 3n - 6$, $f = 2n - 4$.*

2.2. Dual graph

Suppose there are two graphs X_1 and X_2, X_{1i} is any of the subgraphs of X_1, X_{2i} is a subgraph of X_2 corresponding to X_{1i}, X'_{2i} is the complementary graph of X_{2i}. The Graph X_2 is the dual graph of X_1, if the edges of X_1 and X_2 are one-to-one corresponded, and $R(X'_{2i}) = R(X_2) - N(X_{1i})$, where $R(X) = v - 1$, is the rank of the graph, and $N(X) = e - v + 1$, is the nullity of the graph.

If X_2 is the dual graph of X_1, then X_1 is the dual graph of X_2, and $R(X_1) = N(X_2)$, $N(X_1) = R(X_2)$.

A graph has a dual graph if and only if the graph is a planar graph. A planar graph may have multiple dual graphs. A graph contains as much information as its dual graph.

3. Ecological Applications

3.1. Optimum matching problem

Suppose there are six tree species which are used to produce eight mixing planting plans. The biodiversity conservation benefits are as follows

$$
\begin{bmatrix}
15 & 97 & -4 & 15 & 4 & 39 & 7 & 28 \\
3 & 0 & 46 & -1 & 1 & 0 & 0 & 3 \\
7 & 18 & 8 & 0 & 30 & 1 & 2 & 6 \\
0 & -9 & 2 & 0 & 42 & -1 & 28 & 32 \\
9 & 22 & 34 & 1 & 35 & 0 & 0 & 71 \\
45 & 19 & 8 & 18 & -1 & 20 & 1 & 4
\end{bmatrix}.
$$

We need to calculate a plan by which biodiversity is maximally conserved.

The methods, such as Branch-bound Method, can be used to search the tree.

3.2. *Path optimization problem*

There are several scenic spots in an ecological tourism resort. Roads must be constructed between spots but all roads must not be intersected.

In this problem we need to handle the planarity of graph. It is a problem of planar graph.

Algorithms of Graphs

1. Algorithms of Tree

1.1. *DFS algorithm*

DFS (Depth First Search) algorithm is used to obtain a tree from a graph. It is one of the most important algorithms in graph theory.

The procedures of DFS algorithm are as follows.

A graph X is stored using Adjacency Vertex Listing. The ID number of starting vertex to be searched is 1. If T is the set of edges on the tree (k is the sequence number), B is the set of edges that are not on the tree, v is the vertex being checked, w is the vertex to be checked, and num(i) is the ID number of each vertex, then

(1) Let $v = 1, k = 1, j = 1$, num(1) = 1.
(2) Search the incidence edge that is not yet checked.

Take the first edge of v, which has not yet been checked, and set it to be (v, w). Then, reach the vertex w from this edge. The direction of the edge (v, w) is from v to w. Return to (3).

If such an edge has not been found after each of the incident edges of v has been checked, return to (4).
(3) If w is the vertex that has not yet been visited (i.e., num(w) has not yet been determined), then send the edge (v, w) into T, and let $v = w$, $k = k + 1$, and num(w) = k.

If w is the vertex that has been visited (i.e., num(w) $\neq$ 0), then send the edge (v, w) into B, return to the vertex v, and let $j = j + 1$, and return to (2).

(4) Determine the edge (u, v) that orients to vertex v in T. Find out this edge and return to vertex u, let $v = u$, and return to (2). If there does not exists such an edge, terminate the calculation.

The following are Java codes, DFS, for DFS algorithm:

```
//DFS algorithm to obtain a tree in a graph.
/*n: number of vertice; r[n], p[n][]: one-dim array R and
two-dim matrix P in Adjacency Vertice Listing; sets t and b
(t: the edges on the tree; b: the edges not belonging to the
tree) are stored in Two Linear Arrays because all edges in
these two sets are oriented edges. t1[], b1[]: start vertices;
t2[], b2[]: end vertice; num[]: DFS labels of vertice. */
public class DFS {
static int num[],t1[],t2[],b1[],b2[],l,k;
public static void main(String[] args){
int i,j,v,e;
if (args.length!=4) System.out.println("You must input the
names of tables in the database. For example, you may type
the following in the command window: java DFS dfs1 dfs2 21 20,
where dfs1, dfs2 the names of tables, 21 is the number of
vertice, i.e., number of columns of table dfs1 (one-dim array R
in Adjacency Vertice Listing), 20 is the number of columns of
table dfs2 (two-dim matrix P in Adjacency Vertice Listing).");
String tablename1=args[0];
String tablename2=args[1];
v=Integer.valueOf(args[2]).intValue();
e=Integer.valueOf(args[3]).intValue();
readDatabase                                readdata1=new
readDatabase("dataBase",tablename1,v);
readDatabase                                readdata2=new
readDatabase("dataBase",tablename2,e);
int p[][]=new int[v+1][e+1];
int r[]=new int[v+1];
for(i=1;i<=v;i++)
r[i]=(Integer.valueOf(readdata1.data[1][i])).intValue();
for(i=1;i<=v;i++)
for(j=1;j<=e;j++)
p[i][j]=(Integer.valueOf(readdata2.data[i][j])).intValue();
dfs(v,e,r,p); }
public static void dfs(int n, int e, int r[], int p[][]) {
int i,j,v,w,m,s,ss;
num=new int[n+1];
t1=new int[n+1];
```

```
t2=new int[n+1];
b1=new int[n*e+1];
b2=new int[n*e+1];
k=l=v=num[1]=1;
for(i=2;i<=n;i++) num[i]=0;
loopa: do {
ss=r[v];
loopb: do {
for(i=1;i<=ss;i++)
if (p[v][i]!=0) {
w=p[v][i];
p[v][i]=0;
s=r[w];
for(j=1;j<=s;j++)
if (p[w][j]==v) break;
p[w][j]=0;
if (num[w]==0) {
t1[k]=v;
t2[k]=w;
k++;
num[w]=k;
v=w;
continue loopa; }
else {
b1[l]=v;
b2[l]=w;
l++;
continue loopb; } }
if (num[v]==1) break loopa;
else {
m=num[v]-1;
v=t1[m];
continue loopa; } }
while(n>0); }
while(n>0);
System.out.print("DFS labesl of vertice (num[i]): ");
for(i=1;i<=n;i++)
System.out.print(num[i]+" ");
System.out.println();
System.out.print("Start vertice of the edges on the tree
(t1[i]): ");
for(i=1;i<=k-1;i++)
System.out.print(t1[i]+" ");
System.out.println();
System.out.print("Ent vertice of the edges on the tree
```

```
(t2[i]): ");
    for(i=1;i<=k-1;i++)
    System.out.print(t2[i]+" ");
    System.out.println();
    System.out.print("Start  vertice  of  the  edges  not  on  the
tree  (b1[i]): ");
    for(i=1;i<=l-1;i++)
    System.out.print(b1[i]+" ");
    System.out.println();
    System.out.print("End vertice of the edges not on the tree
(b2[i]): ");
    for(i=1;i<=l-1;i++)
    System.out.print(b2[i]+" ");
    System.out.println(); }
    }
```

1.2. *Minty's algorithm*

Minty's algorithm (Minty, 1965) can be used to obtain all trees in a graph. Suppose that an arbitrary edge of a graph X is e_i. Classify all trees into two categories based on e_i, in which a category contains e_i and another does not contain e_i. Find out two subgraphs X_1 and X_2 from X, where e_i is added in X_1 and eliminated in X_2. Every tree in X_1 is added with e_i, which forms the first category of trees in X, and all trees in X_2 belong to the second category of trees in X. Choose another edge and repeat the above procedures to get two subgraphs from X_1 and X_2 respectively. By further repeating the procedures, two new subgraphs can be obtained each time. If the graph becomes a loop, then delete this subgraph. After removing all edges, all edges of the subgraph constitutes a tree. All trees are obtained after every subgraph is handled.

Chan *et al.* (1982) has made a revision on Minty's algorithm. Its Java codes, Minty, are as follows:

```
//Minty algorithm to obtain all trees in a graph.
/*v: number of vertice; e: number of edges; Graph is stored
in Two Linear Arrays. d1[], d2[]: two vertice of an edge are
stored in the two arrays; All vertice in the graph must be
numbered sequentially (from 1 to v); tree[]: a resultant tree
is stored in this array; edge[i]=0, means edge ei is not in the
graph; edge[i]=1, means edge ei is in the graph; edge[i]=-1,
means edge ei is in the graph and is labeled; v[i]=0, means the
vertex i is not in the connected component that composed of
```

labeled edges; v[i]=k, means the vertex i is in the connected component k that composed of labeled edges. */

```java
public class Minty {
public static void main(String[] args){
int i,e;
if (args.length!=1)
System.out.println("You must input the name of table in the database. For example, you may type the following in the command window: java Minty minty, where minty is the name of table.");
String tablename=args[0];
readDatabase                         readdata=new
readDatabase("dataBase",tablename,2);
e=readdata.m;
int a[]=new int[e+1];
int b[]=new int[e+1];
for(i=1;i<=e;i++) {
a[i]=(Integer.valueOf(readdata.data[i][1])).intValue();
b[i]=(Integer.valueOf(readdata.data[i][2])).intValue(); }
Minty(e,a,b); }
public static void Minty(int e, int d1[], int d2[]) {
int i,j,k,l=0,n,m,t,v1=0,v2=0,s,f,v;
n=0;
for(i=1;i<=e;i++) {
if (d1[i]>n) n=d1[i];
if (d2[i]>n) n=d2[i]; }
int edge[]=new int[e+1];
int vmem[][]=new int[n*e+1][n+1];
int emem[][]=new int[n*e+1][e+1];
int tree[]=new int[n+1];
int vert[]=new int[n+1];
for(i=1;i<=e;i++) edge[i]=1;
for(i=1;i<=n;i++) vert[i]=0;
k=f=1;
s=0;
loop: do {
for(j=1;j<=e;j++)
if (edge[j]==1) {
l=j;
edge[j]=m=0;
for(i=1;i<=e;i++) if (edge[i]!=0) m++;
if (m>=(n-1)) {
for(i=1;i<=e;i++) emem[f][i]=edge[i];
for(i=1;i<=n;i++) vmem[f][i]=vert[i];
f++; }
edge[l]=-1;
```

```
v1=d1[1];
v2=d2[1];
if (vert[v1]==0) {
if (vert[v2]==0) {
vert[v1]=k;
vert[v2]=k;
k++;
continue loop; }
else vert[v1]=vert[v2]; }
else {
if (vert[v2]==0) vert[v2]=vert[v1];
else {
l=vert[v1];
m=vert[v2];
if ((l-m)==0) break;
if ((l-m)>0) {
t=m;
m=l;
l=t; }
for(i=1;i<=n;i++) {
if ((vert[i]-m)<0) continue;
else if ((vert[i]-m)==0) {
vert[i]=l;
continue; }
else vert[i]-=1; }
k--; } }
for(i=1;i<=n;i++)
if (vert[i]!=1) continue loop;
s++;
l=1;
for(i=1;i<=e;i++)
if (edge[i]==(-1)) {
tree[l]=i;
l++; }
System.out.println("All edges of tree "+s+":");
for(i=1;i<=l-1;i++)
System.out.print(tree[i]+" ");
System.out.println("\ n"); }
if (f==1) break;
f--;
for(i=1;i<=e;i++) edge[i]=emem[f][i];
k=0;
for(i=1;i<=n;i++) {
vert[i]=vmem[f][i];
if (vmem[f][i]<k) continue;
k=vmem[f][i]; }
```

```
k++; }
while (n>0); }
}
```

1.3.   *Chan' algorithm*

Chan's algorithm is a method of topological analysis. It was used to calculate the directed trees of a graph (Chan *et al.*, 1982):

(1)   Generate the incidence matrix A for describing a directed graph X.
(2)   For the directed graph X with n vertices and e edges, calculate combinatorial number by taking $n - 1$ elements from e elements, i.e., the upper limit J of trees.
(3)   Generate a submatrix A_r from A, i.e., the columns corresponding to r tree branches. Check if A_r is a directed tree. i.e., (i) if it contains some circuit, and (ii) if the outdegree of every vertex is one; or else return to (5).
(4)   Output the admittance product of tree branches.
(5)   Calculate if it reaches the upper limit J. If it has reached J, terminate calculation; or else return to (3).

The procedures for calculating 2-tree is similar to the above.

The following are some of the Java codes for Chan's algorithm, which should be revised:

```
//Chan's algorithm for oriented tree.
/*v: number of vertice; e: number of edges; Graph is stored in
Two Linear Arrays. d1[], d2[]: two vertice of an edge are stored
in the two arrays; All vertice in the graph must be numbered
sequentially (from 1 to v); tree[]: a resultant tree is stored
in this array; edge[i]=0, means edge ei is not in the graph;
edge[i]=1, means edge ei is in the graph; edge[i]=-1, means
edge ei is in the graph and is labeled; v[i]=0, means the vertex
i is not in the connected component that composed of labeled
edges; v[i]=k, means the vertex i is in the connected component
k that composed of labeled edges.*/
public class orienTree {
public orienTree(int k0, int, k1, int nt[], int nf[], int c[],
int a[][]) {
private int i, j, k, n, m;
e=d1.length;
v=0;
```

```
for(i=1;i<=e;i++) {
if (d1[i]>v) v=d1[i];
if (d2[i]>v) v=d2[i]; }
private int edge[]=new int[e+1];
private int vmem[][]=new int[n+1][n+1];
private int emem[][]=new int[n+1][e+1];
private int tree[]=new int[1000];
private int v[]=new int[n+1];
amtx(k0, k1, nt, nf, a);
System.out.println('''');
for(i=1;i<=k1;i++)
System.out.println(i+'' ''+nt[i]+'' ''+nf[i]);
System.out.println('''');
for(i=1;i<=k0;i++) {
for(j=1;j<=k1;j++)
System.out.print(a[i][j]+'' '');
System.out.println(''''); }
k2=k0-1;
ncomb(k2, k1, ncb);
System.out.println(ncb);
for(k=1;k<=ncb;k++) {
combn(k2, k1, k, c);
atmtx(k0, k1, ncb, a, c, at, k2);
tstdr(k0, k2, at, dir);
if (dir<=0) continue;
l=1;
for(i=1;i<=k1;i++) {
if (c[i]<=0) continue;
iy[l]=i;
l++; }
System.out.println('''');
for(j=1;j<=k2;j++)
System.out.print(iy[j]+'' ''); }
k2=k0-2;
ncomb(k2, k1, ncb);
System.out.println(ncb);
for(i=1;i<=ncb;i++) {
combn(k2, k1, k3, c);
atmtx(k0, k1, ncb, a, c, at, k2);
tstdr(k0, k2, at, dir);
if (dir>0) continue;
tstlp(k0, k2, ncn, at, itst);
if (itst<=0) continue;
dcode(at, k0, k2, c, sink1, part1, ll);
l=1;
for(j=1;j<=k1;j++) {
```

```
if (c[j]<=0) contine;
iy[l]=j;
l++; }
System.out.println('''');
for(j=1;j<=k2;j++)
System.out.print(iy[j]+'' '');
System.out.println('''');
for(j=1;j<=ll;j++)
System.out.print(''Delta(''+sink1+'',''+part1[i]+'')''); } }

public void combn(int k1, int k2, int k3, int c[], int nf[]) {
if ((k3-1)<0) return;
else if ((k3-1)==0) {
for(i=1;i<=k1;i++) c[i]=0;
for(i=1;i<=k2;i++) c[i]=1;
return; }
i=k1;
if (c[i]<=0) {
do
i--;
while (c[i]<=0);
c[i]=0;
i++;
c[i]=1;
return; }
l=1;
do {
i--;
if (c[i]>0) l++;
else break; }
while (l!=0);
do {
i--;
if (c[i]<=0) continue;
else break; }
while (l!=0);
c[i]=0;
i++;
k4=i+1;
k5=k4+1;
for(i=1;i<=k4;i++)
c[i]=1;
for(i=k5;i<=k1;i++)
c[i]=0; }

public void tstlp(int k0, int k2, int ncb, int at[][], int itst) {
}
```

2.   Connectedness Testing Algorithm

2.1.   *Connectedness of graph*

The principle of testing connectedness through vertex-fusion algorithm is that for the adjacency matrix of a graph, starting from a vertex, fuse all its adjacent vertices, and then fuse the new added vertices adjacent to it, until no new adjacent vertex is added. A connected component is thus obtained. While fusing the vertices v_i and v_j, add row j of adjacency matrix to row i, and add column j to column i, and then delete row j and column j. In this way, all connected components can be obtained.

The following are Java codes, connDis, for calculating connectedness of a graph, which can output the number of connected components and all vertices of each connected component:

```
//Calculate connectivity of graph. d[][]: adjancy matrix; g[]:
if vertex i belongs to jth connected component then g[i]=j.
  public class connDis {
  public static void main(String[] args){
  int i,j,v;
  if (args.length!=2)
  System.out.println("You must input the name of table in the
database. For example, you may type the following in the command
window: java connDis 6 conndis, where conndis is the name of
table, 6 means the number of vertice.");
  v=Integer.valueOf(args[0]).intValue();
  String tablename=args[1];
  readDatabase                                      readdata=new
readDatabase("dataBase",tablename,v);
  v=readdata.m;
  int a[][]=new int[v+1][v+1];
  for(i=1;i<=v;i++)
  for(j=1;j<=v;j++)
  a[i][j]=(Integer.valueOf(readdata.data[i][j])).intValue();
  conn(v,a); }
  public static int[] conn(int v, int d[][]) {
  int i,j,k,s,t,m,h,a1,a2;
  int g[]=new int[v+1];
  for(i=1;i<=v;i++) g[i]=0;
  s=1;
  t=1;
  loop: do {
  g[t]=s;
  a1=0;
```

```
for(j=1;j<=v;j++) a1+=d[j][t];
m=t+1;
do {
for(h=m;h<=v;h++) {
if ((d[t][h]==0) | (g[h]!=0)) continue;
g[h]=s;
for(i=m;i<=v;i++) {
if (d[i][h]==0) continue;
d[i][t]=1;
d[t][i]=1; } }
a2=0;
for(j=1;j<=v;j++) a2+=d[j][t];
if ((a1-a2)<0) {
a1=a2;
continue; }
if ((a1-a2)>=0) break; }
while(v>0);
for(k=m;k<=v;k++) {
if (g[k]==0) break;
if (k==v) break loop; }
t=k;
s++;
} while(v>0);
System.out.println("Vertice  and  belonged  connectivity
component:");
for(i=1;i<=v;i++)
System.out.print(g[i]+" ");
System.out.println();
System.out.println("Number of connectivity components in
the graph:"+s);
return g; }
}
```

2.2. Block, cut vertex, bridge

Theorem 1. *Two edges belong to the same block, if and only if there exists a circuit that contains the two edges.*

Theorem 2. *A connected graph X is a block, if and only if for any three vertices u, v and w in X, there exists a path from u to w and the path does not contain v.*

Theorem 2 reveals that there is no bottleneck in a block. The vertex v is a bottleneck if any path from u to w must go through v. In this case v is a cut vertex.

According to Theorem 2, the blocks, cut vertices and bridges of a graph can be obtained by calculating the fundamental circuit set of the graph. The DFS algorithm (Tarjan, 1972), for calculating blocks, cut vertices and bridges, are based on Theorem 2.

The following are Java codes, cutVertex, for calculating cut vertices:

```
//Calculate cutvertex of graph.
/*In the vector returned if cut[i]=i then vertex i is the
cutvertex. n: number of vertice; et: number of toward edges;
eb: number of backward edges; sets t and b (t: the edges on the
tree; b: the edges not belonging to the tree) are stored in Two
Linear Arrays because all edges in these two sets are oriented
edges. t1[1—et], b1[1—eb]: start vertices; t2[1—et], b2[1—eb]:
end vertice; t1[1—et], t2[1—et]: set of toward edges; b1[1—eb],
b2[1—eb]: set of backward edges. num[1—n]: DFS labels of
vertice.*/
public class cutVertex {
static int cut[];
public static void main(String[] args){
int i,j,v,e,et,eb;
if (args.length!=4) System.out.println("You must input the
names of tables in the database. For example, you may type the
following in the command window: java cutVertex cutvertex1
cutvertex2 21 20, where cutvertex1, cutvertex2 the names of
tables, 21 is the number of vertice, i.e., number of columns
of table cutvertex1 (1-dim array R in Adjacency Vertice Listing),
20 is the number of columns of table cutvertex2 (2-dim matrix P
in Adjacency Vertice Listing).");
String tablename1=args[0];
String tablename2=args[1];
v=Integer.valueOf(args[2]).intValue();
e=Integer.valueOf(args[3]).intValue();
readDatabase                                  readdata1=new
readDatabase("dataBase",tablename1,v);
readDatabase                                  readdata2=new
readDatabase("dataBase",tablename2,e);
int p[][]=new int[v+1][e+1];
int r[]=new int[v+1];
int num[]=new int[v+1];
int t1[]=new int[v*e+1];
int t2[]=new int[v*e+1];
int b1[]=new int[v*e+1];
int b2[]=new int[v*e+1];
for(i=1;i<=v;i++)
```

```
r[i]=(Integer.valueOf(readdata1.data[1][i])).intValue();
for(i=1;i<=v;i++)
for(j=1;j<=e;j++)
p[i][j]=(Integer.valueOf(readdata2.data[i][j])).intValue();
DFS dfs=new DFS();
dfs.dfs(v,e,r,p);
et=dfs.k-1;
eb=dfs.l-1;
for(i=1;i<=v;i++) num[i]=dfs.num[i];
for(i=1;i<=et;i++) {
t1[i]=dfs.t1[i];
t2[i]=dfs.t2[i]; }
for(i=1;i<=eb;i++) {
b1[i]=dfs.b1[i];
b2[i]=dfs.b2[i]; }
cutVertex(v,et,eb,t1,t2,b1,b2,num); }
public static int[] cutVertex(int n, int et, int eb, int t1[],
int t2[], int b1[], int b2[], int num[]) {
int i,v1,v2,s;
int lw[]=new int[n+1];
cut=new int[n+1];
for(i=1;i<=n;i++) {
cut[i]=0;
lw[i]=num[i]; }
for(i=1;i<=eb;i++) {
v1=b1[i];
v2=b2[i];
if (lw[v1]>=num[v2]) lw[v1]=num[v2]; }
for(i=1;i<=et;i++) {
v1=t1[et-i+1];
v2=t2[et-i+1];
if (lw[v2]<=lw[v1]) lw[v1]=lw[v2]; }
s=0;
for(i=1;i<=et;i++) {
v1=t1[i];
v2=t2[i];
if (v1==1) s++;
if ((lw[v2]>=num[v1]) & (v1!=1)) cut[v1]=v1; }
if (s>=2) cut[1]=1;
System.out.println();
System.out.println("Cutvertex set: ");
for(i=1;i<=n;i++)
System.out.print(cut[i]+" ");
return cut; }
}
```

2.3.   *Vertex connectivity*

Connectivity is a property for connectedness of graph (Chan *et al.*, 1982). By removing a minimal set of vertices such that the graph is an unconnected or trivial graph, then the number of vertices in this set is called connectivity of the graph (or vertex connectivity).

If the connectivity of a graph is K, the minimal degree (the degree of the vertex that has minimal associated edges) is d, and the number of vertices and edges are n and e respectively, then $1 \leq K \leq d \leq 2e/n$.

For regular graph (all vertices have the same degree), $d = 2e/n$. If $K = d$, then the graph has a maximal connectivity. The connectivity of tree is minimal.

The following are Java codes, connCal, for calculating vertex connectivity:

```
/*Calculate between-vertex connectivity. Between-vertex
connectivity is returned as a matrix c[n][n]. Graph is stored
in adjacency matrix d[n][n].*/
  public class connCal {
  public static void main(String[] args){
  int i,j,v;
  if (args.length!=2)
  System.out.println("You must input the name of table in the
database. For example, you may type the following in the command
window: java connCal 21 conncal, where conncal is the name of
table, 21 means the number of vertice.");
  v=Integer.valueOf(args[0]).intValue();
  String tablename=args[1];
  readDatabase                                     readdata=new
readDatabase("dataBase",tablename,v);
  v=readdata.m;
  int a[][]=new int[v+1][v+1];
  for(i=1;i<=v;i++)
  for(j=1;j<=v;j++)
  a[i][j]=(Integer.valueOf(readdata.data[i][j])).intValue();
  conn(v,a);  }
  public static int[][] conn(int v, int d[][]) {
  int i,j,k,r,m,s,t,a,b,conn;
  int c[][]=new int[v+1][v+1];
  int t1[]=new int[v+1];
  int t2[]=new int[v+1];
  int v1[]=new int[v+1];
  int v2[]=new int[v+1];
```

```
int u[]=new int[v+1];
for(s=1;s<=v-1;s++)
for(t=s+1;t<=v;t++) {
conn=0;
loopa: do {
i=s;
for(j=1;j<=v;j++) {
t1[j]=t2[j]=0;
v1[j]=v2[j]=0;
u[j]=0; }
loopb: do {
b=1;
for(k=1;k<=v;k++)
if ((d[i][k]==1) & (d[k][i]==0)) b=0;
for(j=1;j<=v;j++) {
if ((j==s) | (v1[j]<0) | (v2[j]<0) | (d[j][i]==0)) continue;
if (d[i][j]==0) {
if (((v1[i]>0) & (t1[i]==0)) | (v1[i]<0))
{if ((v1[j]>0) & (t1[j]==1)) {
v2[j]=-i;
if (j==t) break loopb;
continue; } } }
if (v1[j]!=0) continue;
if ((i==s) | ((v1[i]>0) & (t1[i]==0)) & (b==1) | (v1[i]<0)) {
v1[j]=i;
if (j==t) break loopb;
continue; }
if ((v1[i]>0) & (t1[i]==1)) {
u[j]=1;
v1[j]=i;
if (j==t) break loopb;
continue; } }
t1[i]=1;
if (v2[i]<0) t2[i]=1;
for(k=1;k<=v;k++)
if ((v1[k]!=0) & (t1[k]==0) | (v2[k]!=0) &
(t2[k]==0)) {
i=k;
continue loopb; }
break loopa; }
while(v>0);
conn++;
r=t;
do {
if (v1[r]>=0) {
m=v1[r];
```

```
d[r] [m]=0;
a=0; }
else {
m=-v1[r];
d[m] [r]=1;
a=1; }
do {
if (a!=0) {
d[m] [r]=1;
a=1; }
if (m==s) continue loopa;
r=m;
if (u[r]!=0) m=-v2[r];
else break; }
while(v>0); }
while(v>0); }
while(v>0);
c[s] [t]=conn; }
for(s=1;s<=v-1;s++) {
c[s] [s]=0;
for(t=s+1;t<=v;t++)
c[t] [s]=c[s] [t]; }
c[v] [v]=0;
System.out.println("Between-vertex connectivity:");
for(i=1;i<=v;i++) {
for(j=1;j<=v;j++)
System.out.print(c[i] [j]+" ");
System.out.println(); }
return c; }
}
```

3. Planarity Testing Algorithm

The computational complexity is huge when we test planarity of a graph using computer, based on Kuratowski Theorem. To test a graph with n vertices, the computational complexity will not be less than $O(n^6)$ (Hopcroft and Tarjan, 1974).

Hopcroft and Tarjan's Algorithm can be used to test planarity of a graph. The computational complexity of the algorithm may reach $O(n)$.

The procedures of Hopcroft and Tarjan's Algorithm are:

(1) Pre-test the graph. If the graph is unconnected, then handle every connected components; if connected components are separable, then

discompose the graph as blocks at the cut vertices and, test the planarity of every block. A block is a planar graph if the number of its edges is greater than $3v - 6$. The graph is a planar graph if all blocks are planar subgraphs.

(2) Calculate a circuit, and discompose all edges, but not the circuit, into paths. First, suppose the starting point of the circuit is the vertex 1. Starting from vertex 1, we go through vertices by new vertex ID numbers based on DFS algorithm, until the first vertex v_0 that has return edge. The first return edge is $(v, 1)$ if edges are numbered with the sequence set by the algorithm. A circuit C, starting from vertex 1, going through the DFS tree to v, and then returning to vertex 1 from $(v, 1)$, is thus obtained.

Second, starting from vertex v, record a path when it meets a return edge based on DFS algorithm. Some paths will return an edge and some consists of the final return edge together with some edges of DFS tree. The starting point of the new generated path is just the starting point of the return edge of last path. The total number of paths is $e - v$.

(3) Embed those paths into the circuit one by one.

The planarity of a graph can also be tested by testing the circuit matrix of the graph (Dunn and Chan, 1968). The algorithm is:

(1) Rearrange all rows of the circuit matrix B, such that the number of 1's in each row is not less than that in the last row. Matrix B_1 is thus obtained.

(2) Generate a n-dimensional column vector E from B_1, where n is the number of rows of B_1, and kth element of E corresponds to the kth row of B_1, and the value is the number of 1's in kth row.

(3) Yield a n-dimensional circuit combinatorial vector C_1, starting from $C = (1\,1, \ldots, 1\,0\,0, \ldots, 0)$, where the number of 1's in C is $e - v + 2$, and the sequential number represented by 1's in C_1 is the sequential number of the circuit joining in the circuit combinatorial vector. Test it by starting from C.

(4) Calculate matrix product $s = C_1 E$. If $s < 2e$, return to (3); if $s = 2e$, return to (5), and if $s > 2e$, return to (6).

(5) Generate matrix B, where its $e - v + 2$ rows are the rows of B_1 represented by element 1 in C_1. If and only if there are two 1's in each

column of B_2, the graph is a planar graph therefore the calculation can be terminated; or else return to (3).

(6) Check whether the number of zeros contained between two adjacent 1's in C_1 is not greater than one. If it is true, then the graph is a planar graph, thereby terminate the calculation; or else return to (3).

4.   Construct Graph from Interrelationship Data

It is possible to construct a graph from between-vertex interrelationship data. Interrelationship may be linear correlation, net correlation, Jaccard coefficient, point correlation, and rank correlation, etc (Zhang, 2007a,b).

Linear correlation between vertices i and j is

$$r_{ij} = \sum_{k=1}^{n}((a_{ik} - a_{ibar_r})(a_{jk} - a_{jbar})) \Bigg/ \left[\sum_{k=1}^{n}(a_{ik} - a_{ibar})^2 \sum_{k=1}^{n}(a_{jk} - a_{jbar})^2\right]^{1/2}$$

where, $-1 \leq r_{ij} \leq 1$, a_{ik} and a_{jk} are kth sample of sampling set of vertices i and j respectively, a_{ibar} and a_{jbar} are means of a_{ik} and a_{jk} respectively, and n is the number of elements in the sampling set.

Net correlation between vertices i and j is

$$R_{ij} = -r_{ij}/(r_{ii}^* r_{jj})^{1/2}$$

where $-1 \leq R_{ij} \leq 1$, and r_{ij} is the element in inverse matrix of linear correlation matrix.

For two vertices, if the correlation is statistically significant, then there is an edge between the two vertices, and the correlation coefficient can be taken as the weight of the edge.

The t-test values of linear correlation and net correlation are:

$$t = r_{ij}/[(1 - r_{ij}^2)/(n - 2)]^{1/2}$$

and

$$t = r_{ij}/[(1 - r_{ij}^2)/(n - m)]^{1/2}$$

respectively, where m is the number of vertices. If $t > t_\alpha$, then the correlation is statistically significant.

Jaccard correlation is:

$$J_{ij} = (e - (b_i + b_j))/(e + b_i + b_j),$$

where $0 \le J_{ij} \le 1$, b_i is the number of element pairs of non-zero numbers for i sampling set but not for j; b_j is the number of element pairs of non-zero numbers for j sampling set but not for i; and e is the number of element pairs of non-zero numbers for both i sampling set and j.

Point correlation is

$$d_{ij} = (ad - bc)/((a + b)(c + d)(a + c)(b + d))^{1/2}$$

where $-1 \le d_{ij} \le 1$, both sampling set i and sampling set j take values zero or one. The number of both sampling sets i and j which takes value 0 is a, b is the number of sampling set i takes value 0 and sampling set j takes value 1, c is number of sampling set i that takes 1 and sampling set j takes 0, and d is number of both sampling sets i and j take value 1.

χ^2-test value of point correlation is:

$$\chi^2 = n(ad - bc)^2/[(a + b)(c + d)(a + c)(b + d)].$$

If $\chi^2 > \chi_\alpha^2$, then point correlation is statistically significant.

Spearman rank correlation is:

$$r_{ij} = 1 - 6^* \sum d^2/[n(n^2 - 1)]$$

where $-1 \le r_{ij} \le 1$, $d = r(i) - r(j)$, and $r(i)$ and $r(j)$ are rank of an element in the sampling set of i and j, from the smaller to the larger values in n elements.

The Java codes, correeCoeff, for calculating correlation and making statistic test are as follows:

```
/*m: number of vertice; n: number of samplings; sig: significance
level, e.g., 0.01, 0.05, etc.; a[][]: matrix of sampling data. */
public class correeCoeff {
public static void main(String[] args){
int i,j,type,m,n;
double sig;
if (args.length!=4)
System.out.println("You must input the type of method (0: Pearson
```

linear correlation, 1: pure linear correlation, 2: Jaccard
coefficient, 3:point correlation, 4: Spearman rank correlation),
number of samplings, and the name of table in the database. For
example, you may type the following in the command window: java
correCoeff 0 54 0.01 correcoeff, where correcoeff is the name
of table, 0 means Pearson linear correlation, 54 means the number
of samplings, i.e., number of columns of table correcoeff in
the database, 0.01 means significance level, e.g., 0.01, 0.05,
etc.");

```
type=Integer.valueOf(args[0]).intValue();
n=Integer.valueOf(args[1]).intValue();
sig=Double.valueOf(args[2]).doubleValue();
String tablename=args[3];
readDatabase readdata=new readDatabase("dataBase",tablename,
n);
m=readdata.m;
double a[][]=new double[m+1][n+1];
for (i=1;i<=m;i++)
for (j=1;j<=n;j++)
a[i][j]=(Double.valueOf(readdata.data[i][j])).doubleValue();
switch (type) {
    case 0: pearsoncorre(m,n,sig,a); break;
    case 1: purecorre(m,n,sig,a); break;
    case 2: bool(type,m,n,a,sig); break;
    case 3: bool(type,m,n,a,sig); break;
    case 4: spearman(m,n,sig,a); break;
} }

public static double[][] pearsoncorre(int m, int n, double sig,
double a[][]) {
int i,j,kk;
double cor,aa,bb,cc,xbar,ybar;
double r[][]=new double[m+1][m+1];
for(i=1;i<=m-1;i++)
for(j=i+1;j<=m;j++) {
cor=aa=bb=cc=xbar=ybar=0;
for(kk=1;kk<=n;kk++) {
xbar+=a[i][kk];
ybar+=a[j][kk]; }
xbar/=n;
ybar/=n;
for(kk=1;kk<=n;kk++) {
aa+=(a[i][kk]-xbar)*(a[j][kk]-ybar);
bb+=Math.pow(a[i][kk]-xbar,2);
cc+=Math.pow(a[j][kk]-ybar,2); }
r[i][j]=aa/Math.sqrt(bb*cc); }
```

```java
for(i=1;i<=m;i++)
r[i][i]=1;
for(i=1;i<=m-1;i++)
for(j=i+1;j<=m;j++)
r[j][i]=r[i][j];
System.out.println("Pearson linear correlation coefficients:");
for(i=1;i<=m;i++) {
for(j=1;j<=m;j++)
System.out.print(r[i][j]+" ");
System.out.println("\n"); }
System.out.println("Vertex pairs with statistically significant
Pearson linear correlation:");
for(i=1;i<=m-1;i++) {
for(j=i+1;j<=m;j++) {
aa=r[i][j]/Math.sqrt((1-Math.pow(r[i][j],2))/(n-2));
if (ttest(aa, n-2)<sig)
System.out.print("("+i+","+j+")"+"("+r[i][j] +") "); }
System.out.println("\n"); }
return r; }

public static double[][] purecorre(int m, int n, double sig,
double a[][]) {
int i,j,kk,ii,jj;
double cor,aa,bb,cc,xbar,ybar;
double r[][]=new double[m+1][m+1];
double rr[][]=new double[m+1][m+1];
for(i=1;i<=m-1;i++)
for(j=i+1;j<=m;j++) {
cor=aa=bb=cc=xbar=ybar=0;
for(kk=1;kk<=n;kk++) {
xbar+=a[i][kk];
ybar+=a[j][kk]; }
xbar/=n;
ybar/=n;
for(kk=1;kk<=n;kk++) {
aa+=(a[i][kk]-xbar)*(a[j][kk]-ybar);
bb+=Math.pow(a[i][kk]-xbar,2);
cc+=Math.pow(a[j][kk]-ybar,2); }
r[i][j]=aa/Math.sqrt(bb*cc); }
for(i=1;i<=m;i++)
r[i][i]=1;
for(i=1;i<=m-1;i++)
for(j=i+1;j<=m;j++)
r[j][i]=r[i][j];
for(kk=1;kk<=m;kk++) {
for(ii=1;ii<=m;ii++)
```

```
for(jj=1;jj<=m;jj++)
if ((ii!=kk) & (jj!=kk))
r[ii][jj]-=r[ii][kk]*r[kk][jj]/r[kk][kk];
for(jj=1;jj<=m;jj++)
if (jj!=kk) {
r[kk][jj]/=r[kk][kk];
r[jj][kk]/=-r[kk][kk]; }
r[kk][kk]=1/r[kk][kk]; }
for(kk=1;kk<=m;kk++)
for(ii=1;ii<=m;ii++)
rr[kk][ii]=-r[kk][ii]/Math.sqrt(r[kk][kk]*r[ii][ii]);
System.out.println("Pure linear correlation coefficients:");
for(i=1;i<=m;i++) {
for(j=1;j<-m;j++)
System.out.print(rr[i][j]+" ");
System.out.println("\n"); }
System.out.println("Vertex pairs with statistically significant
net linear correlation:");
for(i=1;i<=m-1;i++) {
for(j=i+1;j<=m;j++) {
aa=rr[i][j]/Math.sqrt((1-Math.pow(rr[i][j],2))/(n-(m-1)-1));
if (ttest(aa, n-(m-1)-1)<sig)
System.out.print("("+i+","+j+")"+"("+rr[i][j] +") "); }
System.out.println("\n"); }
return rr; }

public static double[][] bool(int type, int m, int n, double
a[][],double sig) {
int i,j,kk;
double cor,aa,bb,cc,dd;
double r[][]=new double[m+1][m+1];
double chi[][]=new double[m+1][m+1];
for(i=1;i<=m-1;i++)
for(j=i+1;j<=m;j++) {
aa=bb=cc=dd=0;
for(kk=1;kk<=n;kk++) {
if((Math.abs(a[i][kk])<=1e-08)&&(Math.abs(a[j][kk])<=1e-08))
aa++;
if((Math.abs(a[i][kk])<=1e-08)&&(Math.abs(a[j][kk])>1e-08))
bb++;
if((Math.abs(a[i][kk])>1e-08)&&(Math.abs(a[j][kk])<=1e-08))
cc++;
if((Math.abs(a[i][kk])>1e-08) && (Math.abs(a[j][kk])>1e-08))
dd++; }
chi[i][j]=n*Math.pow(aa*dd-bb*cc,2)/((aa+bb)*(cc+dd)*(aa+cc)*
(bb+dd));
```

```
if (type==2) r[i][j]=(dd-(cc+bb))/(dd+cc+bb);
else            if              (type==3)
r[i][j]=(aa*dd-bb*cc)/Math.sqrt((aa+bb)*(cc+dd)*(aa+cc)*(bb+
dd)); }
for(i=1;i<=m-1;i++)
for(j=i+1;j<=m;j++)
r[j][i]=r[i][j];
for(i=1;i<=m;i++)
r[i][i]=1;
if (type==2) System.out.println("Jaccard coefficients:");
else if (type == 3) System.out.println("Point correlation
coefficients:");
for(i=1;i<=m;i++) {
for(j=1;j<=m;j++)
System.out.print(r[i][j]+" ");
System.out.println("\n"); }
if (type==3) {
System.out.println("Vertex pairs with statistically significant
point correlation:");
for(i=1;i<=m-1;i++) {
for(j=i+1;j<=m;j++) {
if (chitest(n-1,chi[i][j])<sig)
System.out.print("("+i+","+j+") "+"("+r[i][j] +") "); }
System.out.println("\n"); } }
return r; }

public static double[][] spearman(int m, int n, double sig,
double a[][]) {
int i,j,ii,jj,nx,ny,ntie;
double rs,aa;
int rx[]=new int[n+1];
int ry[]=new int[n+1];
double r[][]=new double[m+1][m+1];
double x[]=new double[n+1];
double y[]=new double[n+1];
for(ii=1;ii<=m-1;ii++)
for(jj=ii+1;jj<=m;jj++) {
for(j=1;j<=n;j++) {
nx=ny=1;
for(i=1;i<=n;i++) {
if (a[ii][i]<a[ii][j]) nx++;
if (a[jj][i]<a[jj][j]) ny++; }
rx[j]=nx;
ry[j]=ny; }
for(j=1;j<=n;j++) {
if (rx[j]==n+1) continue;
```

```
nx=rx[j];
ntie=-1;
for(i=1;i<=n;i++) {
if (rx[i]!=nx) continue;
ntie++;
x[i]=rx[i];
rx[i]=0; }
for(i=1;i<=n;i++) {
if (rx[i]!=0) continue;
x[i]+=(ntie*0.5);
rx[i]=n+1; } }
for(j=1;j<=n;j++) {
if (ry[j]==n+1) continue;
ny=ry[j];
ntie=-1;
for(i=1;i<=n;i++) {
if (ry[i]!=ny) continue;
ntie++;
y[i]=ry[i];
ry[i]=0; }
for(i=1;i<=n;i++) {
if (ry[i]!=0) continue;
y[i]+=(ntie*0.5);
ry[i]=n+1; } }
rs=0;
for(i=1;i<=n;i++)
rs+=Math.pow(x[i]-y[i],2);
r[ii][jj]=1-((6*rs)/(n*(Math.pow(n,2)-1))); }
for(i=1;i<=m-1;i++)
for(j=i+1;j<=m;j++)
r[j][i]=r[i][j];
for(i=1;i<=m;i++)
r[i][i]=1;
System.out.println("Spearman correlation coefficients:");
for(i=1;i<=m;i++) {
for(j=1;j<=m;j++)
System.out.print(r[i][j]+" ");
System.out.println("\n"); }
System.out.println("Vertex pairs with statistically significant
Spearman correlation:");
for(i=1;i<=m-1;i++) {
for(j=i+1;j<=m;j++) {
aa=r[i][j]/Math.sqrt((1-Math.pow(r[i][j],2))/(n-2));
if (ttest(aa, n-2)<sig)
System.out.print("("+i+","+j+")"+"("+r[i][j] +") "); }
```

```
System.out.println("\n"); }
return r; }
public static double ttest(double tv, int df) {
double yyy,sss,rrr,zzz,jjj,kkk,lll,tvv;
double aa=1;
yyy=1;
tvv=tv*tv;
if (tvv<1) {
sss=df;
rrr=yyy;
zzz=1/tvv; }
else {
sss=yyy;
rrr=df;
zzz=tvv; }
jjj=2.0/9/sss;
kkk=2.0/9/rrr;
lll=Math.abs((1-kkk)*Math.pow(zzz,1/3.0)-1+jjj)/Math.sqrt(kk
k*Math.pow(zzz,2/3.0)+jjj);
if (rrr<4) lll*=1+0.08+Math.pow(lll,4)/Math.pow(rrr,3);
aa=0.5/Math.pow(1+lll*(0.196854+lll*(0.115194+lll*(0.000344+
lll*0.019527))),4);
if (tvv<1) aa=1-aa;
return aa; }
public static double chitest(int df,double chi) {
int i,r;
double kp,jjj,lll,mm,aa;
r=1;
i=df;
while (i>=2) {
r*=i;
i-=2;}
kp=Math.pow(chi,(int)((df+1)/2.0))*Math.exp(-chi/2)/r;
if (df/2.0==(double)(df/2)) jjj=1;
else jjj=Math.sqrt(2/chi/3.1415926);
lll=1;
mm=1;
while (mm>=1e-05) {
df+=2;
mm*=chi/df;
lll+=mm; }
aa=jjj*kp*lll;
return aa; }
}
```

5.   Computer Generation of Graph

A graph can be generated by a well designed computer algorithm. The graph is recorded using Two Linear Array, as indicated in the following example. The directed graph in Fig. 1 can be expressed in Two Linear Array as:

$$S_1 = (v_1, v_1, v_1, v_2, v_2, v_2, v_3, v_3, v_5, v_6),$$
$$S_2 = (v_2, v_4, v_3, v_2, v_3, v_4, v_4, v_4, v_5, v_6),$$
$$S = (1, 1, -1, 3, 1, 1, 2, -2, 3, 0).$$

If the graph in Fig. 1 is an undirected graph, then its Two Linear Array expression is:

$$S_1 = (v_1, v_1, v_1, v_2, v_2, v_2, v_3, v_4, v_5, v_6),$$
$$S_2 = (v_2, v_3, v_4, v_2, v_3, v_4, v_4, v_3, v_5, v_6),$$
$$S = (1, -1, 1, 3, 1, 1, 2, 2, 5, 4).$$

In this storage way, 1 means there is only one edge between two vertices and the edge is a positive-directed edge; -1 means there is only one edge between two vertices and the edge is a negative-directed edge; 2 means two parallel edges; 3 means self-loop; 4 means isolated vertex; 5 means the self-loop of an isolated vertex.

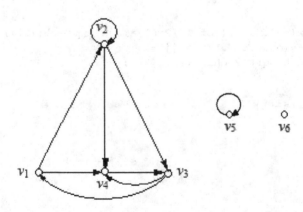

Figure 1.   A directed graph.

The following are Java codes, netGenerator, for generating a graph:

```
//Loading example: java netGenerator netgenerator
//where netgenerator is the name of table in the database
''dataBase''
/*Graph is stored in Two Arrays Listing. s1[1—e]: from-vertex;
s2[1—e]: to-vertex; tt[1—e]: 1 if it is a forward edge; —1
is for backward edge; 4 if s1[i]=s2[i], i.e., the vertex is
an isolated vertex with no self-loop; 2: parallel edges; 3:
self-loop for non-isolated vertex; 5: an isolated vertex with
self-loop. */
public class netGenerator {
public static void main(String[] args){
String tablename=args[0];
readDatabase                              readdata=new
readDatabase("dataBase",tablename, 3);
int mm=readdata.m;
String s1[]=new String[mm+1];
String s2[]=new String[mm+1];
int s[]=new int[mm+1];
for (int i=1;i<=mm;i++) {
for (int i=1;i<=mm;i++) {
s1[i]=readdata.data[i][1];
s2[i]=readdata.data[i][2];
s[i]=(Integer.valueOf(readdata.data[i][3])).intValue(); }
new GraphicsFrame(new NetGraph(s1,s2,s),"Black and blue edges
are forward and backward edges respectively. Self-loop is
labeled by semicircle on the vertex. Parallel edges are labeled
by between-vertex parallel lines.").resize(720,550);
} }
```

The following are the classes NetGraph, NetVertex, NetEdge, and Net-Panel, used by class netGenerator:

```
import java.util.*;
import java.awt.*;
import java.applet.Applet;
import java.awt.event.*;
public class NetGraph extends Applet implements ActionListener {
Button close;
NetPanel panel;
Panel controlPanel;
String s[][], c[];
int e, v, m;
public int t[];
```

```
public NetGraph(String s1[], String s2[], int tt[]) {
e=s1.length;
s=new String[5][e+1];
c=new String[2000];
t=new int[e+1];
for(int i=1;i<=e;i++) {
s[1][i]=s1[i];
s[2][i]=s2[i];
t[i]=tt[i]; }
v=1;
c[1]=s[1][1];
for(int i=1;i<=2;i++)
for(int j=1;j<=e;j++) {
m=0;
for(int k=1;k<=v;k++)
if (!(s[i][j].equals(c[k]))) m++;
if (m==v) {
v++;
c[v]=s[i][j]; } }
begin(); }

public void begin() {
close=new Button("Close");
setLayout(new BorderLayout());
panel=new NetPanel();
add("Center", panel);
controlPanel=new Panel();
add("South",controlPanel);
controlPanel.add(close); close.addActionListener(this);
for(int k=1;k<=e;k++)
panel.addEdge(s[1][k],s[2][k],t[k]);
Dimension d=getSize();
resize(d.width-15, d.height-15);
setLocation(20,20);
validate();
show(); }

public void init() {
}

public void paint(Graphics g) {
repaint(); }
```

```
public void destroy() {
remove(panel);
remove(controlPanel); }
public void actionPerformed(ActionEvent e) {
Object src=e.getSource();
if (src==close) {
this.hide();
this.getParent().hide();
System.exit(0); }
return; }

public String getAppletInfo() {
return "Ecological Network Generator"; }
}

class NetVertex {
double x,y,w,h;
boolean fixed;
String lab; }

class NetEdge {
int from,to,type; }

class NetPanel extends Panel implements MouseListener,
MouseMotionListener {
int nvertice,nedges;
NetVertex vertice[]=new NetVertex[2000];
NetEdge edges[]=new NetEdge[50000];
NetVertex pick;
boolean pickfixed;
Image offscreen;
Dimension offscreensize;
Graphics offgraphics;
Color posColor=Color.black;
Color negColor=Color.blue;

public NetPanel() {
addMouseListener(this);
addMouseMotionListener(this); }

public int findVertex(String lab) {
for(int i=0;i<nvertice;i++)
```

```
if (vertice[i].lab.equals(lab)) return i;
return addVertex(lab); }

public int addVertex(String lab) {
NetVertex v=new NetVertex();
v.x=100+350*Math.random();
v.y=100+350*Math.random();
v.lab=lab;
vertice[nvertice]=v;
return nvertice++; }

public void addEdge(String from, String to, int type) {
NetEdge e=new NetEdge();
e.from=findVertex(from);
e.to=findVertex(to);
e.type=type;
edges[nedges++]=e; }

public void paintVertex(Graphics g, NetVertex v, FontMetrics
mtr) {
int x=(int)v.x;
int y=(int)v.y;
g.setColor(Color.white);
int w=mtr.stringWidth(v.lab)+10;
int h=mtr.getHeight()+4;
v.w=w;
v.h=h;
g.fillRect(x-w/2,y-h/2,w,h);
g.setColor(Color.black);
g.drawRect(x-w/2,y-h/2,w-1,h-1);
g.drawString(v.lab,x-(w-10)/2,(y-(h-4)/2)+mtr.getAscent());
}

public void update(Graphics g) {
Dimension d=getSize();
if ((offscreen==null) || (d.width!=offscreensize.width) ||
(d.height!=offscreensize.height)) {
offscreen=createImage(d.width, d.height);
offscreensize=d;
offgraphics=offscreen.getGraphics();
offgraphics.setFont(getFont()); }
offgraphics.setColor(getBackground());
offgraphics.fillRect(0,0,d.width,d.height);
```

```
for(int i=0;i<nedges;i++) {
NetEdge e=edges[i];
int x1=(int)vertice[e.from].x;
int y1=(int)vertice[e.from].y;
int x2=(int)vertice[e.to].x;
int y2=(int)vertice[e.to].y;
if ((e.type==1) | (e.type==-1)) {
if (e.type==1) offgraphics.setColor(posColor);
else offgraphics.setColor(negColor);
offgraphics.drawLine(x1,y1,x2,y2); }
if (e.type==2) {
offgraphics.setColor(posColor);
offgraphics.drawLine(x1,y1-5,x2,y2-5);

offgraphics.setColor(negColor);
offgraphics.drawLine(x1,y1+5,x2,y2+5); }
if ((e.type==3) | (e.type==5)) {
int w=(int)vertice[e.from].w;
int h=(int)vertice[e.from].h;
int rad=w-1;
offgraphics.setColor(posColor);
offgraphics.drawArc(x1-(int)(0.5*w),y1-h,rad,rad,0,360); } }
FontMetrics mtr=offgraphics.getFontMetrics();
for(int i=0;i<nvertice;i++) paintVertex(offgraphics,vertice
[i],mtr);
g.drawImage(offscreen,0,0,null); }

public void paint(Graphics g) {
repaint(); }

public void mousePressed(MouseEvent e) {
double bestdist=Double.MAX_VALUE;
int x=e.getX();
int y=e.getY();
for(int i=0;i<nvertice;i++) {
NetVertex v=vertice[i];
double dist=(v.x-x)*(v.x-x)+(v.y-y)*(v.y-y);
if (dist<bestdist) {
pick=v;
bestdist=dist; } }
pickfixed=pick.fixed;
pick.fixed=true;
pick.x=x;
```

```
pick.y=y;
repaint();
e.consume(); }

public void mouseReleased(MouseEvent e) {
pick.x=e.getX();
pick.y=e.getY();
pick.fixed=pickfixed;
pick=null;
repaint();
e.consume(); }

public void mouseDragged(MouseEvent e) {
pick.x=e.getX();
pick.y=e.getY();
repaint();
e.consume(); }

public void mouseClicked(MouseEvent e) {}

public void mouseEntered(MouseEvent e) {}

public void mouseExited(MouseEvent e) {}

public void mouseMoved(MouseEvent e) {}
}
```

The following is the class GraphicsFrame, used by class netGenerator:

```
import java.awt.*;
import java.applet.*;
public class GraphicsFrame extends Frame {
public GraphicsFrame(Applet applet) {
this.resize(600,400);
add(applet);
setVisible(true); }

public GraphicsFrame(Applet applet, String str) {
this.resize(600,400);
this.setTitle(str);
add(applet);
setVisible(true); }
}
```

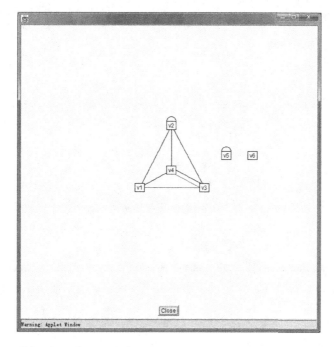

Figure 2.   The graph of Fig. 1 generated by computer algorithm.

Figure 2 is the graph of Fig. 1, generated by the preceding algorithm.

6.   Ecological Applications

For an ecological graph (network), the connectivity is a connectedness property of the graph. A larger connectivity means a better connectedness of the graph. The graph (network) thus has a larger redundancy, a stronger disturbance resistance, and a stronger stability.

CHAPTER 6

Directed Graphs

A directed graph can be represented by its fundamental graph together with arrows on it.

There are many examples for the directed graph, for example, network, flow chart, finite-state machine (Fig. 1), etc.

1. Directed Graph

Corresponding to a directed graph X, we may generate a graph X' based on the same vertex set with X, such that corresponding to every edge of X, X' has an edge that shares the same endpoints. The resulting graph is called the fundamental graph of X. Many concepts on directed graph are defined based on fundamental graph. For example, a directed graph is called connected if its fundamental graph is connected. Path, circuit, tree, cutset, etc., can all be defined according to fundamental graph.

Given an arbitrary graph X, assign an order for the endpoints of every linkage. An edge is thus determined and a directed graph is obtained. The resulting directed graph is called an oriented graph. The oriented graph of a complete graph is called a tournament graph.

A directed graph is called strict if it does not have any loop and any two edges do not share the same direction and endpoints.

If there exists a directed path (u, v) in X, vertex v is called reachable in X starting from vertex u. Two vertices are called strongly connected (bilaterally connected; Fig. 2), if they are reachable to each other. Strongly connectedness is an equivalence relation with respect to the vertex set of X.

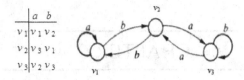

Figure 1.   A finite-state machine and its illustration.

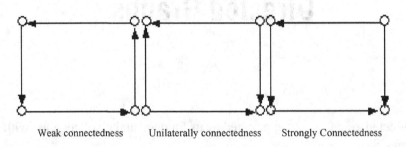

| | Weak connectedness | Unilaterally connectedness | Strongly Connectedness |

Figure 2.   Several types of connectedness.

The directed subgraphs $X(V_1), X(V_2), \ldots, X(V_m)$, induced by a classification $(V_1, V_2, \ldots, V_m)$ of vertex set $V(X)$ which is based on the relation of strongly connectedness, are called strongly connected components of X. A directed graph X is strongly connected if it has one and only one strongly connected component.

There are also unilaterally connectedness, weak connectedness, etc (Fig. 2). Suppose X is a directed graph, u and v are any two vertices in X. If

(1) v is reachable from u, or u is reachable from v, then X is unilaterally connected.
(2) For every pair of vertices (u, v), at least v is reachable from u, or u is reachable from v, then X is weak connected.
(3) For every pair of vertices (u, v), there exists a vertex w, such that u and v are reachable from w, then X is pseudo-strongly connected.

Theorem 1. *X is strongly connected, if and only if X is connected and every block of X is strongly connected.*

Theorem 2. *A directed graph X is strongly connected, if and only if every edge of X is in a directed circuit.*

Theorem 3. *A directed graph X is unilaterally connected, if and only if every edge of X is in a directed path.*

Equivalently, there are the following:

A directed graph X is strongly connected, if and only if there exists a circuit in X that passes through every vertex at least once.

A directed graph X is unilaterally connected, if and only if there exists a path in X that passes through every vertex at least once.

A strongly connected graph must be a unilaterally connected graph. A unilaterally connected graph must be a weak connected graph.

2. Directed Euler Graph

In a directed graph X, if there exists a directed closed chain that passes through all the edges of X, then the directed closed chain is called a Euler directed chain.

If there exists a Euler directed chain in a directed graph X, then X is called a Euler directed graph. If a directed graph is a Euler directed graph, then it must be strongly connected.

The directed graph X in Fig. 3 is strongly connected, where $e_1 e_2 e_4 e_3 e_5 e_7 e_6$, is a Euler directed chain from and to v_1.

A directed graph is a Euler directed graph if and only if for all vertices in X, $d^+(v) = d^-(v)$, $v \in V(X)$.

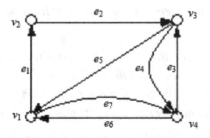

Figure 3.   A Euler directed graph.

3.   Directed Path

The Hamilton path of a directed graph X refers to the directed path that contains every vertex of X. Any tournament graph contains the Hamilton path.

Theorem 4 (Roy, 1967; Gallai, 1968). *Any directed graph X contains a directed path with length $\chi - 1$, where χ is the chromatic number of vertices.*

Theorem 5. *There always exists an independent set S in a directed graph X which does not have any loop, such that for every vertex that is not present in S, a vertex is reachable by starting from a vertex in S and going through a directed path with at most length two.*

Theorem 6. *Suppose G is a subgraph of connected directed graph, and v is an arbitrary vertex in G. If $d^+(v) = d^-(v)$, i.e., the outdegree of v is equal to its indegree, and $d(v) = 2$, then G is a directed circuit.*

Theorem 7. *Suppose X is a connected directed graph. For any $v \in V(X)$, if $d^+(v) = 1$ (or $d^-(v) = 1$), then there is one and only one directed circuit in X.*

Theorem 8. *Suppose G is the subgraph of a connected directed graph X; if*

(1) $d^+(u) = 1$, $d^-(u) = 0$, $d^+(v) = 0$, $d^-(v) = 1$, $u, v \in V(G)$;
(2) *For any $w \in V(G)$, $d^+(w) = d^-(w) = 1$,*

then G is a (u, v) directed path.

Theorem 9. *If there is no directed circuit in a directed graph X, then there is at least one vertex with zero outdegree and at least one vertex with zero indegree.*

Theorem 10. *There is no directed circuit in a directed graph if and only if all edges of X can be eliminated through w procedure. The w procedure means that suppose v is a vertex of a directed graph X, if $d^+(v) = 0$ (or $d^-(v) = 0$), then eliminate vertex v and its associative edges.*

Theorem 11. *A directed graph X contains a directed circuit if and only if there exists a subgraph in which every vertex satisfies the condition, $d^+(v) > 0$, $d(v) > 0$.*

4. Directed Tree

In a directed graph X, if there exists a vertex r, such that any vertex in X is reachable by starting from r, then r is called the root of X.

Theorem 12. *A directed graph X has a root if and only if X is pseudo-strongly connected.*

Suppose X is a rooted directed graph. If the fundamental graph of X is a tree, then X is called a directed tree.

For any vertex v in a directed tree, there exists a directed path starting from the root and ending at v. Therefore the directed tree is also called an outgoing tree. If every edge of a directed tree is in reverse direction, then the tree is called ingoing tree.

Theorem 13. *Suppose X is a directed graph with the number of vertices $n > 1$. The following statements are equivalent:*

(1) *X is a pseudo-strongly connected graph without any circuit.*
(2) *X is pseudo-strongly connected and it has $n - 1$ arcs.*
(3) *X is a rooted tree.*
(4) *There exists a vertex such that for every other vertex there is one and only one directed path starting from the root r.*
(5) *X is pseudo-strongly connected, but it does not hold if any of the edges is eliminated from X.*
(6) *X is pseudo-strongly connected and there exists a vertex r such that $d^-(r) = 0$, $d^+(v) = 1$, $v \neq r$.*
(7) *X does not have any circuit and there exists a vertex r such that $d^-(r) = 0$, $d^+(v) = 1$, $v \neq r$.*

We thus know that X contains the directed spanning tree if and only if X is pseudo-strongly connected.

In a directed tree, the vertices with zero outdegree are called leaves and the other vertices are called branching vertices. The length of a path from the root r to a vertex v, is called the number of layers of vertex v (Fig. 4).

When we specify an order for vertices of every layer in a directed tree, then the directed tree is called an ordered tree. In general the order within the same layer is from left to right, or else we specify the directed tree by defining the order of edges.

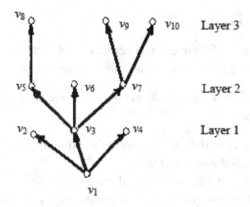

Figure 4.   A three-layer directed tree.

5.   Signal Flow Chart

As early as in 1942, Shannon found that the Gramer Law for solving linear equations can be described using the directed graph (Shannon, 1942). In 1953, Mason proposed the signal flow chart method for linear systems (Mason, 1953).

Signal flow chart is a directed graph with weighted vertices and weighted edges. In a signal flow chart, between-variable relationship in linear equations is represented by the graphical structure. Vertices denote variables and edges denote coefficients. Because there are correspondence relationship between matrix and linear equations, there are also correspondence relationship between the algebraic transformation of linear equations and the transformation of graph. Therefore, the signal flow chart is a method for solving linear equations.

It is not necessary to establish the mathematical model of a linear system in the signal flow chart method. It can also be used to handle the problems that coefficients are symbols. Due to the simple and flexible representation, signal flow chart is widely used in linear systems (linear networks).

For a linear ecosystem, its input and output satisfy some algebraic equations. Signal flow chart of the ecosystem can be constructed directly from properties of the chart, and we may thus write out the solution of the corresponding equations based on the signal flow chart and some rules.

The equations of a linear ecosystem at some time (Zhang, 2007b), for example, are as follows:

$$dx_1/dt = a_{11}x_1 + a_{12}x_2 + a_{13}x_3 + \cdots + a_{1n}x_n = b_1$$
$$dx_2/dt = a_{21}x_1 + a_{22}x_2 + a_{23}x_3 + \cdots + a_{2n}x_n = b_2$$
$$\vdots$$
$$dx_n/dt = a_{n1}x_1 + a_{n2}x_2 + a_{n3}x_3 + \cdots + a_{nn}x_n = b_n$$

It is obvious that they are linear equations and the coefficient matrix is $A = (a_{ij})_{n \times n}$. For professional convenience, the signal flow chart can be called information flow chart in ecological applications.

5.1. Mason's flow chart

Transform the following algebraic equations

$$a_{11}x_1 + a_{12}x_2 + a_{13}x_3 + \cdots + a_{1n}x_n = b_1$$
$$a_{21}x_1 + a_{22}x_2 + a_{23}x_3 + \cdots + a_{2n}x_n = b_2$$
$$\vdots$$
$$a_{n1}x_1 + a_{n2}x_2 + a_{n3}x_3 + \cdots + a_{nn}x_n = b_n$$

to

$$x_1 = d_1 + c_{11}x_1 + c_{12}x_2 + c_{13}x_3 + \cdots + c_{1n}x_n$$
$$x_2 = d_2 + c_{21}x_1 + c_{22}x_2 + c_{23}x_3 + \cdots + c_{2n}x_n$$
$$\vdots$$
$$x_n = d_n + c_{n1}x_1 + c_{n2}x_2 + c_{n3}x_3 + \cdots + c_{nn}x_n$$

where the term $x_j = c_{ji}x_i$, corresponds to the directed edge from x_i to x_j, and the edge weight is c_{ji}. The Mason's flow chart of the algebraic equations can thus be constructed. For example, if $n = 4$, the Mason's flow chart is indicated in Fig. 5.

5.2. Coates's flow chart

In the Coates flow chart, a matrix is represented by a weighted directed graph. For a non-zero element a_{ij}, draw a directed edge from v_i to v_j, and

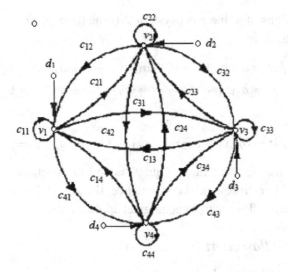

Figure 5.   A Mason's flow chart.

if the edge weight is a_{ij}, then the matrix $A = (a_{ij})_{n \times n}$ corresponds to a weighted directed graph with n vertices, i.e., Coates flow chart.

For example, the following matrix

$$A = \begin{bmatrix} 0 & 0 & 3 & 1 \\ 0 & 2 & 6 & 3 \\ 7 & 1 & 0 & 0 \\ 0 & 5 & 1 & 0 \end{bmatrix}$$

Its Coates flow chart is Fig. 6.

The properties of a matrix may be studied by using its Coates flow chart.

5.3.   *Calculation of signal flow chart*

To calculate signal flow chart, the following rules are formulated:

(1) **Addition.** The m edges with the same direction between two vertices can be replaced with one edge with the same direction. The gain of the edge is the weight sum of the m edges.

(2) **Multiplication.** The $m + 1$ vertices $v_1, v_2, \ldots, v_m, v_{m+1}$ can be replaced with the edge $e_{1\,m+1}$, if there exist linked edges e_{12},

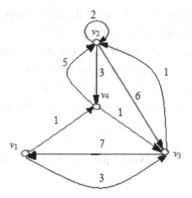

Figure 6. A Coates flow chart.

$e_{23}, \ldots, e_{m\,m+1}$ with the same direction. The gain of the edge $e_{1\,m+1}$ is the product of weights of the edges $e_{12}, e_{23}, \ldots, e_{m\,m+1}$.

(3) **Vertex elimination.** In general, for the directed edges (v_1, v), $(v_2, v), \ldots, (v_m, v)$ that end with v, and the directed edges (v, w_1), $(v, w_2), \ldots, (v, w_n)$ that start from v, if their gains are $a_1, a_2, \ldots, a_m$, and $b_1, b_2, \ldots, b_n$, respectively, then the vertex v can be eliminated and a directed edge will be achieved

$$(v_1, w_1), (v_2, w_1), \ldots, (v_m, w_1),$$
$$(v_1, w_2), (v_2, w_2), \ldots, (v_m, w_2),$$
$$\vdots$$
$$(v_1, w_n), (v_2, w_n), \ldots, (v_m, w_n).$$

The gains are

$$a_1 b_1, a_2 b_1, \ldots, a_m b_1,$$
$$a_1 b_2, a_2 b_2, \ldots, a_m b_2,$$
$$\vdots$$
$$a_1 b_n, a_2 b_n, \ldots, a_m b_n$$

respectively.

(4) **Self-loop elimination.** In general, for the vertex v with a loop and m directed edges ending at v, if we divide the gains of the m directed

edges by $(1-$ the gain of the loop), then the loop can be eliminated. The gain of the directed edge starting from v is kept constant.

(5) **Reversal.** Reversal refers to the change of directed edge (v_j, v_i) to the directed edge (v_i, v_j) with respect to the source vertex v_j, and to change the edge weight a_{ij} to $1/a_{ij}$. The directed edge (v_k, v_i) is changed to the directed edge (v_k, v_j), and the edge weight a_{ik} is changed to $-a_{ik}/a_{ij}$.

Reversal is effective only when the staring vertex of the directed edge (v_i, v_j) is a source vertex.

Mason's Theorem. *Suppose G is the signal flow chart of equation $AX = B$; $C_1, C_2, \ldots, C_n$, are the circuits of G respectively, and w_i is the gain of circuit L, i.e., the product of weights of all edges of C_i, then*

$$D = \det A = 1 - \sum_{i=0} w_i + \sum_{i,j} w_i w_j - \sum_{i,j,k} w_i w_j w_k - \ldots$$

where $L_i, L_j, L_k, \ldots$, are disjoint for each other, and

$$x_i = (1/D) \sum_{j=1}^{n} b_j T_{jk}$$

$$i = 1, 2, \ldots, n$$

where,

$$T_{jk} = \sum_k w(p_k) \Delta_k$$

$w(p_k)$ is the gain of the path p_k from b_j to x_i, Δ_k is the result of eliminating all terms associated with the vertices in path p_k from $\det A$.

Theorem 14. *The sub-square matrices S_1 and S_2 of matrices M and D, which are generated from the vertices $v_1, v_2, \ldots, v_m$, and the edges $e_1, e_2, \ldots, e_m$, are nonsingular if and only if these edges construct a circuit, or a group of disjoint circuits, where $M = (m_{ij})_{n \times m}$, $m_{ij} = 1$, if $(x_i, x_k) = e_j$; otherwise $m_{ij} = 0$.*

For example, suppose there is an ecosystem with seven variables and the between-variable relationship is as follows:

$$x_2 = c_{11}x_1 + c_{12}x_3$$
$$x_4 = c_{21}x_2 + c_{22}x_3 + c_{23}x_5$$
$$x_6 = c_{31}x_4 + c_{32}x_5 + c_{33}x_7$$

The solution finding process is shown in Fig. 7.

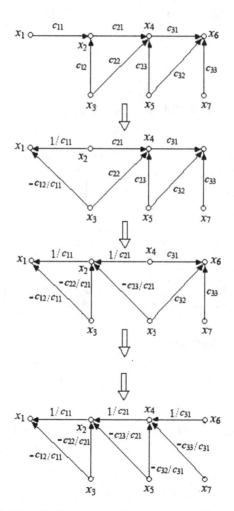

Figure 7.   Solution finding process of an ecosystem with seven variables.

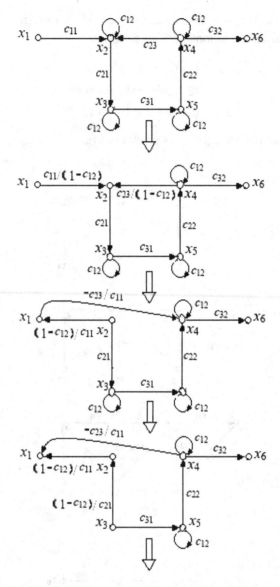

Figure 8.   Solution finding process of an ecosystem with six variables.

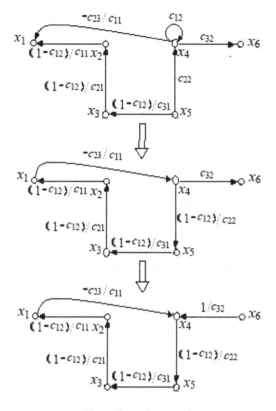

Figure 8. (*Continued*)

Take one more example. Suppose there is an ecosystem with six variables, as indicated in the first diagram of Fig. 8. We want to find the variables' ratio, x_1/x_6.

The solution finding process is shown in Fig. 8. The solution is

$$x_1/x_6 = c_{32}^{-1}((1 - c_{12})^4/(c_{11}c_{21}c_{22}c_{31}) - c_{23}/c_{11}).$$

PART II

Networks

CHAPTER 7

Networks

Graph is the mathematical terminology of network (Bohman, 2009). In many cases, the terminologies of graph and network are used indifferently. Strictly speaking, network is a type of graph. A network X refers to a directed graph having two specific vertex subsets A and B (the fundamental directed graph of X). Both A and B are disjoint and nonempty. The vertices in A are called source vertices and vertices in B are called sink vertices. The vertices not belonging to both A and B are called intermediate vertices. The set of intermediate vertices is denoted by I. For example, in the network of Fig. 1, $A = \{v_1, v_2, v_3\}$, $B = \{v_7, v_8\}$, and $I = \{v_4, v_5, v_6\}$.

In the network, vertices and edges are usually called nodes and arcs respectively. In many cases, network edges are assigned weights and the network is thus a weighted graph.

1. Petrì Net

Graph theory may be used to handle static relationships but not dynamic problems. Petri net is a tool to study complex systems. We may construct the Petri net of a practical problem and make analysis on the Petri net. The dynamic properties and other important information can thus be approached. Petri net is known to offset the shortcoming of a graph theory.

Petri net is a directed graph X

$$X = (V, E),$$

$$V = P \cup T,$$

$$E = I \cup O.$$

123

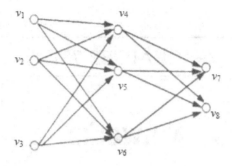

Figure 1.   A network.

where position vertex set $P = \{p_1, p_2, \ldots, p_m\}$, transition vertex set $T = \{t_1, t_2, \ldots, t_m\}$, I is the set of vertices whose directed edges orient to a vertex, O is the set of vertices which are oriented by the directed edges starting from a vertex, and $I, O \subset P$.

A labeled Petri net means that the position vertices in a Petri net are labeled and the number of labels is infinite. Different numbers of labels means different states of the Petri net. If $v \in I$, and the number of labels of v is not less than the number of the edges from v to u, then u will be activated (ignited). After u is activated, the number of labels of the vertex at the input position of u reduces and the reduced number is equal to the number of the edges from v to u. Meanwhile, the number of labels of the vertex w at the output position of u increases and the increased number is equal to the number of the edges from u to w. If position vertex is condition and transition vertex is operation, then the number of labels indicates that whether the condition is ready.

2. Optimization and Dynamic Control of Network

Network is optimized to: (1) search for an optimal search plan, and (2) achieve a topological structure so that the network possesses relative stability. For example, we may need to theoretically prove whether the topological structure of a biological system is optimal.

The goal functions of complex systems are always hard to be analytically expressed. Therefore we hope to achieve an approximation of goal function by nonparametric test or linearization methods.

If goal function is simple, then parameter optimization can be realized through conventional optimization techniques. Otherwise, parameters can be optimized by using sensitivity analysis, statistical experiment, and heuristic methods.

The dynamic control of network means to change the topological structure and key parameters of the network stage by stage so that the goal function of entire network achieves the optimum or suboptimum. Mathematical tools, like dynamic programming, decision-making analysis, game theory, etc., can be used to handle these problems.

2.1. *Shortest path problem (SPP)*

The objective of shortest path problem (SPP) is to find a path with minimal weight sum (shortest path) in a weighted graph. The shortest path refers to the one between two given vertices, or the one between a given vertex to remaining vertices. The weights in weighted graph are distance, cost, flux, etc.

To find between-vertex shortest path of an undirected graph, Dijkstra algorithm can be used (Dijkstra, 1959). First, define the weight matrix $D = (d_{ij})$ of an undirected graph X, where d_{ij} is the weight of the edge e_{ij}. $d_{ij} = 0$, if $i = j$; $d_{ij} > 0$, if there exists an edge e_{ij}, and $d_{ij} = \infty$, if there is not an edge e_{ij}. Suppose the two vertices are A (starting vertex) and B (terminal vertex), then Dijkstra algorithm is as follows:

(1) Label v_A as $v_A = 0$, and the other vertex v_i as $v_i = \infty, i \neq A$.
(2) Label the unlabeled vertex v_j as

$$v_j = (j_{\text{old}}, i_{\text{old}} + d_{ij})$$

(3) Find the minimum of labels and take the label as the fixed label of the vertex; return to (2), until B is labeled. The shortest path and its length are thus achieved.

The following are Java codes, Dijkstra, for Dijkstra algorithm:

```
/*Dijkstra algorithm to calculate the shortest path in a
non-oriented graph.*/
/*v: number of vertice; d[1 - v][1 - v]: the weight matrix of
graph in which d[i][j] = 0, if there is not an edge between
vertice i and j. p[]: vertice in shortest path. */
    public class Dijkstra {
    public static void main(String[] args){
    int i, j, v;
    if (args.length!=2)
    System.out.println("You must input the name of table in the
    database. For example, you may type the following in the
    command window: java Dijkstra 5 dijkstra, where dijkstra is
    the name of table, 5 means the number of  vertice.");
    v = Integer.valueOf(args[0]).intValue();
    String tablename = args[1];
    readDatabase readdata=new readDatabase("dataBase",
tablename,v);
    v = readdata.m;
    double a[][] = new double[v+1][v+1];
    for(i=1;i<=v;i++)
    for(j=1;j<=v;j++)
    a[i][j] = (Double.valueOf(readdata.data[i][j])).
doubleValue();
    Dijkstra(v, a); }
    public static void Dijkstra(int v, double d[][]) {
    int i, j, k, n, c, h;
    double ma,iv,sd,inf = 1e+50;
    double a[] = new double[v+1];
    int w[] = new int[v+1];
    int p[] = new int[v+1];
    int b[] = new int[v+1];
    for(i=1;i<=v;i++)
    for(j=1;j<=v;j++)
    if ((d[i][j]==0) & (i!=j)) d[i][j]=inf;
    for(j=1;j<=v-1;j++)
    for(k=j+1;k<=v;k++) {
    for(i=1;i<=v;i++) {
    p[i] = w[i] = 0;
    a[i] = inf; }
    a[j] = 0;
    w[j] = 1;
    n = j;
    h = 0;
    do {
```

```
ma = inf;
for(i=1;i<=v;i++) {
if (w[i]==1) continue;
iv = d[n][i]+a[n];
if (iv<a[i]) {
a[i] = iv;
b[i] = n; }
if (a[i]>ma) continue;
ma = a[i];
h = i; }
w[h] = 1;
if (h==k) break;
n = h; }
while(v>0);
sd = a[k];
p[1] = k;
c = k;
for(i=2;i<=v;i++) {
if (c==j) break;
p[i] = b[c];
c = b[c]; }
System.out.println("Shortest path from "+j+" to "+k+":");
for(i=v;i>=1;i--)
if ((p[i]!=0) & (sd!=inf)) {
if (i>1) System.out.print(p[i]+"->");
else System.out.println(p[i]); }
if (sd!=inf) System.out.println("Distance="+sd+"\n");
else System.out.println("No path"+"\n"); } }
}
```

To find the shortest path of an undirected graph, Floyd algorithm can also be used (Floyd, 1962). In the algorithm, each time insert a vertex, and compare the weight sum of the path between any two vertices with the weight sum of the path going through the two vertices and inserted vertex, if the former is larger, then replace the former path with the path having inserted the vertex. Suppose the graph X has v vertices, and the weight matrix is $D = (d_{ij})$, then the algorithm is

(1) Let $k = 1$.
(2) Let $i = 1$.
(3) Calculate

$$d_{ij} = \min(d_{ij}, d_{ik} + d_{kj}), \quad j = 1, 2, \ldots, v.$$

(4) Let $i = i + 1$, if $i \leq v$, then return to (3).

(5) Let $k = k + 1$, if $k \leq v$, then return to (2), or else terminate the calculation.

The following are Java codes, Floyd, for Floyd algorithm:

```
/*Floyd algorithm to calculate the shortest path in an
undirected graph.*/
/*v: number of vertice; d[1 - v][1 - v]: the weight matrix of
graph in which d[i][j] = 1e+50, if there is not an edge between
vertice i and j, and then the matrix of between-vertex shortest
distance. a[1 - v][1 - v]: */
    public class Floyd {
    public static void main(String[] args){
    int i, j, v;
    if (args.length!=2)
    System.out.println("You must input the name of table in the
database. For example, you may type the following in the command
window: java Floyd 5 floyd, where floyd is the name of table,
5 means the number of vertice.");
    v = Integer.valueOf(args[0]).intValue();
    String tablename = args[1];
    readDatabase                                      readdata=new
    readDatabase("dataBase",tablename,v);
    v = readdata.m;
    double a[][] = new double[v+1][v+1];
    for(i=1;i<=v;i++)
    for(j=1;j<=v;j++)
    a[i][j] = (Double.valueOf(readdata.data[i][j])).
doubleValue();
    Floyd(v, a); }
    public static void Floyd(int v, double d[][]) {
    int i, j, k, m, n, u, p, q;
    double c,inf = 1e+50;
    int a[][] = new int[v+1][v+1];
    int b[] = new int[10000];
    int e[] = new int[10000];
    int h[] = new int[10000];
    for(i=1;i<=v;i++)
    for(j=1;j<=v;j++)
    if ((d[i][j]==0) & (i!=j)) d[i][j]=inf;
    for(i=1;i<=v;i++)
    for(j=1;j<=v;j++)
    if (d[i][j]!=inf) a[i][j]=j;
```

```
for(i=1;i<=v;i++)
for(j=1;j<=v;j++)
for(k=1;k<=v;k++) {
c = d[j][i]+d[i][k];
if (c<d[j][k]) {
d[j][k] = c;
a[j][k] = i; } }
for(p=1;p<=v;p++)
for(q=1;q<=v;q++) {
if (p==q) continue;
u = a[p][q];
m = 1;
b[1] = u;
do {
m++;
b[m] = a[b[m - 1]][q];
if (q==b[m]) break;
if (b[m]==b[m - 1]) break;
if (m>v) break; }
while(v>0);
n = 1;
e[1] = u;
do {
n++;
e[n] = a[p][e[n - 1]];
if (p==e[n]) break;
if (e[n]==e[n - 1]) break;
if (n>v) break; }
while(v>0);
for(i=1;i<=m+n-1;i++) {
if (i==1) h[i]=p;
if ((i<=n) & (i>1)) h[i] = e[n-i+1];
if ((i>n) & (i<(m+n-1))) h[i] = b[i - n+1];
if (i==(m+n-1)) h[i]=q;}
System.out.println("Shortest path from vertex "+p+"
to vertex "+q+":");
for(i=1;i<=m+n-1;i++)
if ((h[i]!=0) & (d[p][q]!=inf)) {
if ((h[i]==h[i+1]) & (i<m+n-1)) continue;
if (i<m+n-1) System.out.print(h[i]+"->");
else System.out.println(h[i]); }
if (d[p][q]!=inf)
System.out.println("Distance="+d[p][q]+"\n");
else System.out.println("No path"+"\n"); } }
}
```

Besides the above algorithms, there is also a decision tree algorithm that is used in risk analysis, which can be used to determine an optimal action path. The algorithm is:

(1) Construct the decision tree. Draw a tree based on various possible events and processes. The primary elements in a decision tree are decision nodes, plan nodes, final nodes and paths. A decision node generates plan paths and the number of plan paths is the possible number of action plans. A plan node generates probability paths and the number of probability paths is the number of possible natural states. Final nodes are final nodes in a decision tree and the value attaching to each final node is the benefit of this plan under corresponding state.

(2) After the decision tree is constructed, label all nodes including decision nodes, plan nodes and final nodes, from left to right and from top to bottom.

(3) Determine the occurrence probability of events. Estimate the occurrence probability for every possible event. Label the natural state and occurrence probability on the probability path.

(4) Calculate the benefit expectation. Starting from final nodes, from right to left calculate benefit expectation of each action plan based on benefit and corresponding probability.

(5) The path with maximal benefit expectation is the optimal action path.

The Java codes, deciTree, for decision tree algorithm are:

```
/*Decision tree algorithm for risk analysis.*/
/*v: number of vertice; n: number of vertice in the final
layer; Data are stored in Two Array Listing. d1[1 - e], d2[1 - e]:
start and end vertice of edges; d[1 - e]: probabilities of edges;
If the probability is 1 then it means this vertex is a decision
vertex; b[1 - n]: benefits of vertice in the final layer. Vertice
are sequentially numbered in the graph. */
    public class deciTree {
    public static void main(String[] args){
    int i, j, k, v, e, s=0, n;
    double h;
    if (args.length!=2) System.out.println("You must input the
names of tables in the database. For example, you may type the
following in the command window: java deciTree decitree1
decitree2, where decitree1 and decitree2 are the names of
tables. d1[1 - e], d2[1 - e] and d[1 - e] are stored in decitree1;
```

```
b[1 - n] is stored in decitree2.");
   String tablename1 = args[0];
   String tablename2 = args[1];
   readDatabase                                readdata1=new
   readDatabase("dataBase",tablename1, 3);
   readDatabase                                readdata2=new
   readDatabase("dataBase",tablename2, 1);
   e = readdata1.m;
   n = readdata2.m;
   int d1[] = new int[e+1];
   int d2[] = new int[e+1];
   double d[] = new double[e+1];
   double b[] = new double[n+1];
   for (i=1;i<=e;i++) {
   d1[i] = (Integer.valueOf(readdata1.data[i][1])).
intValue();
   d2[i] = (Integer.valueOf(readdata1.data[i][2])).
intValue();
   d[i] = (Double.valueOf(readdata1.data[i][3])).
doubleValue(); }
   for (i=1;i<=n;i++)
   b[i] = (Double.valueOf(readdata2.data[i][1])).
doubleValue();
   v = 0;
   for(i=1;i<=e;i++) {
   if (d1[i]>v) v = d1[i];
   if (d2[i]>v) v = d2[i]; }
   double r[][] = new double[v+1][v+1];
   double p[] = new double[v+1];
   int x[] = new int[v+1];
   for(i=1;i<=v;i++) {
   for(j=1;j<=v;j++) r[i][j]=0;
   p[i] = 0;
   x[i] = 0; }
   for(i=v-n+1;i<=v;i++) p[i]=b[i+n-v];
   for(i=1;i<=v;i++)
   for(j=1;j<=v;j++)
   for(k=1;k<=e;k++)
   if ((d1[k]==i) & (d2[k]==j)) {
   r[i][j]=d[k];
   break; }
   loopa: for(i=v-n;i>=1;i--)
   loopb: for(j=v;j>=1;j--)
   if (r[i][j]==1) {
   h = p[j];
   x[i] = 1;
```

```
for(k = v;k>=1;k--)
if ((h>p[k]) | (r[i][k]==0)) continue;
else {
h = p[k];
s = k; }
x[s] = 1;
p[i] = h;
continue loopa; }
else if (r[i][j]==0) continue loopb;
else p[i]+=r[i][j]*p[j];
r[1][0] = 1;
System.out.println("Optimal path is: ");
for(i=1;i<=v;i++)
for(j=1;j<=v;j++)
if (r[i][0]==0) continue;
else {for(j=1;j<=v;j++)
if (r[i][j]==0) continue;
else if ((x[i]==0) | (x[j]==0)) continue;
else {r[j][0]=1;
if (r[i][j]!=1) continue;
else System.out.print(i+"-->"+j+"\n\n"); } }
System.out.println("Expected benefit in every node:\n");
for(i=1;i<=v;i++)
System.out.println(i+":"+(int)(p[i]*10000)/10000.00+
"\n"); }
}
```

For example, an ecological risk decision problem involves 15 nodes (1 decision node and 14 natural states) and eight final nodes. Occurrence probabilities and other data are listed as follows:

$$d_1 = (v_1, v_1, v_2, v_2, v_3, v_3, v_4, v_4, v_5, v_5, v_6, v_6, v_7, v_7),$$

$$d_2 = (v_2, v_3, v_4, v_5, v_6, v_7, v_8, v_9, v_{10}, v_{11}, v_{12}, v_{13}, v_{14}, v_{15}),$$

$$d = (1, 1, 0.2, 0.4, 0.3, 0.9, 0.3\ 0.8, 0.6\ 0.2, 0.3\ 0.7, 0.2, 0.7).$$

$$b = (26, 11, 18, 20, 34, 25, 31, 27).$$

Using the above decision tree algorithm, we obtained the optimal decision $v_1 \rightarrow v_3$, and the benefit expectation of every nodes are: 1: 30.9; 2: 9.24; 3: 30.9; 4: 16.6; 5: 14.8; 6: 27.7; 7: 25.09; 8: 26.0; 9: 11.0; 10: 18.0; 11: 20.0; 12: 34.0; 13: 25.0; 14: 31.0; 15: 27.0.

2.2. *Dynamic programming*

The theoretical foundation of dynamic programming is Bellman's principle (Bellman, 1957), i.e., the optimal decision sequence of a multi-phase decision process has such a property, that is, regardless of how the initial phase, the initial state and initial decision are, take the phase and state generated in the first decision as the initial condition, the subsequent decisions must constitute the optimal sequences with respect to the corresponding problem (Norton, 1972; Li *et al.*, 1982).

Dynamic programming can be used to optimize network, for example, the network optimization problem, as indicated in Fig. 2.

The discrete and deterministic dynamic programming, used in network optimization, is described as follows.

First, divide the procedure into n phases, $k = 1, 2, \ldots, n$, and determine the state x_k of phase k, x_k is an initial state of phase k.

Second, determine the decision variables of each phase. Suppose $u_k(x_k)$ is the decision variable of phase k when state is x_k. $u_k(x_k) \in D_k(x_k)$, where $D_k(x_k)$ is the admissible decision set of phase k. The decision function series from phase k to the end point, i.e., the substrategies are:

$$P_{kn} = \{u_k(x_k), u_{k+1}(x_{k+1}), \ldots, u_n(x_n)\}$$

Third, determine the rules for state transition. Given the state variable x_k of phase k, use the decision variable u_k, then the state x_{k+1} of phase $k + 1$ can be determined, i.e., $x_{k+1} = T_k(x_k, u_k)$.

Fouth, define the goal function. The goal function V_{kn} is used to evaluate the goodness of the procedure:

$$V_{kn} = V_{kn}(x_k, u_k, x_{k+1}, \ldots, x_{n+1}), k = 1, 2, \ldots, n,$$

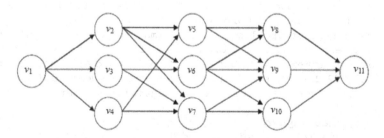

Figure 2.   Discrete and deterministic multi-phase decision problem.

where the optimal value of V_{kn} is the optimal goal function $f_k(x_k)$. Calculation formula of the goal function is:

$$V_{kn} = v_k(x_k, u_k) + V_{k+1n}(x_{k+1}, \ldots, x_{n+1})$$

where $v_k(x_k, u_k)$ is the goal value of phase k. Goal function is the function of initial condition and strategies, so its calculation formula can be written as:

$$V_{kn}(x_k, P_{kn}) = v_k(x_k, u_k) + V_{k+1n}(x_{k+1}, P_{k+1n})$$

where $P_{kn} = \{u_k(x_k), P_{k+1n}(x_{k+1})\}$.

Finally, make reverse sequence optimization:

$$\text{opt}(P_{kn})V_{kn}(x_k, P_{kn}) = \text{opt}(u_k)\{v_k(x_k, u_k) + \text{opt}(P_{k+1n})V_{k+1n}$$
$$k = n, n-1, \ldots, 1$$

$$f_1(x_1) = \text{opt}(P_{1n})V_{1n}(x_1, P_{1n})$$

or

$$f_k(x_k) = \text{opt}(u_k \in D_k(x_k))\{v_k(x_k, u_k) + f_{k+1}(x_{k+1})\}$$
$$k = n, n-1, \ldots, 1$$

$$f_{n+1}(x_{n+1}) = 0$$

where opt is minimization (min) or maximization (max). By starting from $k = n$, calculate ahead until $f_1(x_1)$ is obtained. Then optimal strategy and the optimal value of goal function will be obtained.

The Java codes, dynProg, for the discrete and deterministic dynamic programming, are as follows:

```
/*Dynamic programming algorithm for discrete and deter-
ministic multi-phase problems.*/
   /*type: type of problem, 1 for maximal and 0 for minimal; v:
number of vertice; e: number of edges; s: ID of start vertex, s=1;
t: ID of end vertex, t = v. Data are stored in Two Arrays Listing.
d1[1 - e], d2[1 - e]: start and end vertice of edges; dd[1 - e]:
weights of edges; Vertice are sequentially numbered in the
graph. */
   public class dynProg {
   public static void main(String[] args){
   int i, j, k, z, q=0, v, e, s,t, type;
```

```
double h, w;
if (args.length!=2)
System.out.println("You must input the type of problem
(1 for maximal and 0 for minimal) and the name of table in the
database. For example, you may type the following in the command
window: java dynProg 0 dynprog, where dynprog is the name of
table, 0 means minimal problem.");
type = Integer.valueOf(args[0]).intValue();
String tablename = args[1];
readDatabase                                readdata=new
readDatabase("dataBase",tablename, 3);
e = readdata.m;
int d1[] = new int[e+1];
int d2[] = new int[e+1];
double dd[] = new double[e+1];
for (i=1;i<=e;i++) {
d1[i] = (Integer.valueOf(readdata.data[i][1])).
intValue();
d2[i] = (Integer.valueOf(readdata.data[i][2])).
intValue();
dd[i] = (Double.valueOf(readdata.data[i][3])).
doubleValue(); }
v = 0;
for(i=1;i<=e;i++) {
if (d1[i]>v) v = d1[i];
if (d2[i]>v) v = d2[i]; }
double x[][] = new double[v+1][v+1];
double a[] = new double[v+1];
int p[] = new int[v+1];
int c[] = new int[v+1];
int d[] = new int[v+1];
s = 1;
t = v;
for(i=1;i<=v;i++)
for(j=1;j<=v;j++) x[i][j]=1e+10;
for(i=1;i<=v;i++)
for(j=1;j<=v;j++)
for(k=1;k<=e;k++)
if ((d1[k]==i) & (d2[k]==j)) {
x[i][j]=Math.pow(-1,type)*dd[k];
break; }
for(i=1;i<=v;i++) {
a[i] = 1e+10;
c[i] = 0;
p[i] = 0;}
```

```
a[s] = 0;
c[s] = 1;
z = s;
while (z<=1e+50) {
for(k=1;k<=v;k++)
if ((c[k]==1) | ((a[z]+x[z][k])>=a[k])) continue;
else {
a[k] = a[z]+x[z][k];
p[k] = z; }
h = 1e+10;
for(k=1;k<=v;k++)
if ((a[k]>h) | (c[k]==1)) continue;
else {
h = a[k];
q = k; }
if (h==1e+10) System.exit(0);
z = q;
if (z==t) break;
c[z] = 1; }
w = a[t];
i = 1;
while (i>=1) {
d[i] = t;
if (t==s) break;
t = p[t];
i++; }
System.out.println("Optimal strategy:\n");
while (i>0) {
System.out.print(String.valueOf(d[i])+""+"
("+String.valueOf((int)(Math.pow(-1,type)*a[d[i]]))+")");
if (i==1) break;
System.out.print("-->"+String.valueOf(d[i-1])+""+
"("+String.valueOf((int)(Math.pow(-1,type)*a[d[i-1]]))+")");
if (i==2) break;
System.out.print("-->");
i-=2; }
System.out.println("\n");
if (type==1) System.out.print("Maximum benefit="+
String.valueOf
((int)(-w*100000)/100000.00)+"\n");
else                          System.out.print("Minimum
cost="+String.valueOf((int)(w*100000)/100000.00)+"\n"); }
}
```

For example, an ecological travel network involves 11 scenic spots and 20 paths (Fig. 2). The costs of all paths are as follows:

$$d_1 = (v_1, v_1, v_1, v_2, v_2, v_2, v_3, v_3, v_4, v_4, v_5, v_5, v_6, v_6,$$
$$v_6, v_7, v_7, v_8, v_9, v_{10}),$$

$$d_2 = (v_2, v_3, v_4, v_5, v_6, v_7, v_6, v_7, v_5, v_7, v_8, v_9, v_8, v_9, v_{10}, v_9, v_{10},$$
$$v_{11}, v_{11}, v_{11}),$$

$$dd = (7, 6, 12, 5, 3, 9, 13, 11, 7, 1, 15, 12, 6, 11, 9, 4, 3, 8, 5, 10).$$

We want to find a plan with minimal total cost.

Using the above algorithm, we achieved the optimal path sequence $1 \to 4 \to 7 \to 9 \to 11$; the phase optimums of goal function are $0 \to 12 \to 13 \to 17 \to 22$, and the overall optimum is 22.

2.3. *Maximum flow and minimum cost flow*

A transportation network is a connected directed graph with n vertices. The vertex v_0 is the source vertex, and it does not have any terminal edge. The vertex v_n is the sink vertex, and it does not have any initial edge. The weight c_{ij} of directed edge e_{ij} refers to the maximal transportation capacity of the edge; f_{ij} is the flux of the edge, $0 \le f_{ij} \le c_{ij}$, and the influx sum and outflux sum of vertex v_j are the same:

$$\sum_{i=0}^{n-1} f_{ij} = \sum_{k=1}^{n} f_{jk}, \quad j = 1, 2, \ldots, n-1$$

Therefore, the outflux sum of source vertex is equal to the influx sum of sink vertex.

2.3.1. *Maximum flow*

The form of maximum flow problem is:

$$\max \sum_{j=1}^{n} f_{0j}$$

Max-Flow and Min-Cut Theorem (Ford–Fulkerson Theorem, 1956): In a transportation network, the value of maximum flow is equal to the capacity of a minimum cutting of the network.

Based on Ford–Fulkerson Theorem, Ford and Fulkerson (1957) proposed an algorithm to find maximum flow. The principle of the algorithm is that, starting from a given flow, recursively construct a series of flows with increasing values and terminate at the maximum flow. Once every new flow f is generated, if there exists a f-incremental path g, then find the path g, construct a revised flow f' which is based on g, and take it as the next flow of the series. If there is not a f-incremental path g, terminate calculation and thus, f is the maximum flow.

The main procedures of Ford–Fulkerson algorithm is as follows:

(1) Labeling process.

    (a) Label the source vertex Vs with $(+, +\infty)$, $d_s = +\infty$.

    (b) Choose a labeled vertex x. For all unlabeled adjacent vertices y of x, handle them by the following rules:

        If $yx \in E$, and $f_{yx} > 0$, let $d_y = \min\{f_{yx}, d_x\}$, and label y with (x, d_y).

        If $xy \in E$, and $f_{xy} < C_{xy}$, let $d_y = \min\{C_{xy} - f_{xy}, d_x\}$, and label y with $(x+, d_y)$.

    (c) Repeat (b) until the sink vertex v_t is labeled or no more vertex can be labeled. If v_t is labeled, then there exists an augmenting chain, and return to (2) for adjusting the process; if v_t is not labeled, labeling process is not able to be conducted, then f is the maximum flow.

(2) Adjust process.

    (a) Determine adjust magnitude $d = d_{vt}$, and let $u = v_t$.

    (b) If the vertex u is labeled as $(v+, d_u)$, then replace f_{vu} with $f_{vu} + d$; if the vertex u is labeled as (v, d_u), then replace f_{vu} with $f_{vu} + d$.

    (c) If $v = v_s$, then remove all labels and return to (1) for labeling again; or else let $u = v$, return to (b).

Once the calculation terminates, let the set of labeled vertices be S, then cutset (S, Sc) is the minimum cut, and the maximum flow is $M_f = C(S, Sc)$.

The following are Java codes, maxiFlow, for Ford–Fulkerson algorithm:

```
/*Ford--Fulkerson algorithm for maximum flow.*/
/*v: number of vertice; Data are stored in Two Array Listing.
d1[1 - e], d2[1 - e]: start and end vertice of edges;
   d[1-e]: flow capacity of edges;
Vertice are sequentially numbered in the graph. */
public class maxiFlow {
public static void main(String[] args){
int i,e;
if (args.length!=1)
System.out.println("You must input the name of table in the
   database. For example, you may type the following in the
   command window: java maxiFlow maxiflow, where maxiflow is
   the name of table.");
String tablename = args[0];
readDatabase                                   readdata=new
   readDatabase("dataBase",tablename,3);
e = readdata.m;
int a[] = new int[e+1];
int b[] = new int[e+1];
double c[] = new double[e+1];
for(i=1;i<=e;i++) {
a[i] = (Integer.valueOf(readdata.data[i][1])).intValue();
b[i] = (Integer.valueOf(readdata.data[i][2])).intValue();
c[i] = (Double.valueOf(readdata.data[i][3])).doubleValue(); }
maxiFlow(e, a, b, c); }
public static void maxiFlow(int e, int d1[], int d2[],
double dd[]) {
int i, j, k, v, pr, s;
double mf,dv;
v = 0;
for(i=1;i<=e;i++) {
if (d1[i]>v) v = d1[i];
if (d2[i]>v) v = d2[i]; }
double c[][] = new double[v+1][v+1];
double f[][] = new double[v+1][v+1];
int no[] = new int[v+1];
double d[] = new double[v+1];
for(i=1;i<=v;i++)
for(j=1;j<=v;j++)
for(k=1;k<=e;k++)
if ((d1[k]==i) & (d2[k]==j)) {
c[i][j] = dd[k];
break; }
for(i=1;i<=v;i++)
for(j=1;j<=v;j++) f[i][j]=0;
```

```
for(i=1;i<=v;i++) {
no[i] = 0;
d[i] = 0; }
pr = 1;
while(v>0) {
no[1] = v+1;
d[1] = 1e+30;
while(v>0) {
pr = 1;
for(i=1;i<=v;i++) {
if (no[i]!=0) {
for(j=1;j<=v;j++) {
if ((no[j]==0) & (f[i][j]<c[i][j])) {
no[j] = i;
d[j] = c[i][j] - f[i][j];
pr = 0;
if (d[j]>d[i]) d[j]=d[i]; }
else if ((no[j]==0) & (f[j][i]>0)) {
no[j] = - i;
d[j] = f[j][i];
pr = 0;
if (d[j]>d[i]) d[j]=d[i]; } } } }
if ((no[v]!=0) | (pr!=0)) break; }
if (pr!=0) break;
dv = d[v];
s = v;
while (v>0) {
if (no[s]>0) f[no[s]][s]+=dv;
else if (no[s]<0) f[no[s]][s]-=dv;
if (no[s]==1) {
for(i=1;i<=v;i++) {
no[i] = 0;
d[i] = 0; }
break; }
s = no[s]; } }
mf = 0;
for(j=1;j<=v;j++) mf+=f[1][j];
System.out.println("Maximum flow:");
for(i=1;i<=v;i++) {
for(j=1;j<=v;j++)
System.out.print(f[i][j]+" ");
System.out.println(); }
System.out.println("Maximum flow=:"+mf);
System.out.println("Labels for minimum cut:");
for(i=1;i<=v;i++)
System.out.print(no[i]+" "); }
}
```

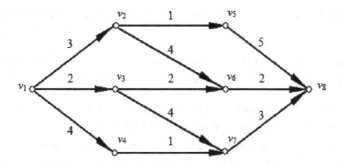

Figure 3. A species invasion network.

For example, in the species invasion network of Fig. 3, there are eight sites and 11 paths. Maximum number of species for every path is indicated in the Fig. 3

The data of the network are as follows:

$$d_1 = (v_1, v_1, v_1, v_2, v_2, v_3, v_3, v_4, v_5, v_6, v_7),$$
$$d_2 = (v_2, v_3, v_4, v_5, v_6, v_6, v_7, v_7, v_8, v_8, v_8),$$
$$dd = (3, 2, 4, 1, 4, 2, 4, 1, 5, 2, 3).$$

The maximum total number matrix was achieved based on Ford–Fulkerson algorithm as the following:

$$
\begin{bmatrix}
0 & 3 & 2 & 1 & 0 & 0 & 0 & 0 \\
0 & 0 & 0 & 0 & 1 & 2 & 0 & 0 \\
0 & 0 & 0 & 0 & 0 & 0 & 2 & 0 \\
0 & 0 & 0 & 0 & 0 & 0 & 1 & 0 \\
0 & 0 & 0 & 0 & 0 & 0 & 0 & 1 \\
0 & 0 & 0 & 0 & 0 & 0 & 0 & 2 \\
0 & 0 & 0 & 0 & 0 & 0 & 0 & 3 \\
0 & 0 & 0 & 0 & 0 & 0 & 0 & 0
\end{bmatrix}
$$

The vertex set of labels is $\{9, 0, 0, 1, 0, 0, 0, 0\}$. The maximum total number is six.

According to max-flow and min-cut theorem, we can obtain the following theorem:

Theorem 1. *Suppose X is a network with v_0 as the source vertex and v_n as the sink vertex, and each edge has a unit capacity, then*

(1) *The value of maximum flow in X is equal to the maximal number of directed (v_0, v_n) paths with unrepeated edges in X.*
(2) *The capacity of minimum cutset in X is equal to the minimal number of such edges, if these edges are eliminated then all directed (v_0, v_n) paths in X will be damaged.*

Theorem 2. *A graph X is k-edge connected if and only if any two different vertices in X are linked by a path with at least k unrepeated edges.*

Theorem 3. *Suppose v_0 and v_n are two vertices of a graph X, then the maximal number of the (v_0, v_n) paths with unrepeated edges is equal to the minimal number of such edges, if these edges are eliminated then all (v_0, v_n) paths in X will be damaged.*

2.3.2. *Minimum cost flow*

Minimum cost flow problem refers to that in the maximum flow problem, set transportation cost a_{ij} on each edge, try to achieve the flux $\sum f_{0j}$ with v_0 as the source vertex and v_n as the sink vertex, and minimize the total cost:

$$\min \sum_{i,j} a_{ij} f_{ij}$$

Given the network $X = (V, E, C)$, take the initial available flow f as the zero flow. The procedures to find minimum cost flow are:

(1) Construct weighted directed graph $X_f = (V, E_f, F)$, for any $e_{ij} \in E$, E_f and F are defined as

$$e_{ij} \in E_f, \quad F(e_{ij}) = b_{ij}, \quad \text{if } f_{ij} = 0;$$

$$e_{ji} \in E_f, F(e_{ji}) = b_{ij}, \quad \text{if } f_{ij} = C_{ij}$$

$$e_{ij} \in E_f, F(e_{ij}) = b_{ij}, e_{ji} \in E_f, F(e_{ji}) = -b_{ij}, \quad \text{if } 0 < f_{ij} < C_{ij}.$$

Return to (2).

(2) Find the shortest path h from source vertex v_s to sink vertex v_t in the weighted directed graph $X_f = (V, E_f, F)$. If there exists a shortest path h, return to (3), or else f is the maximum flow with minimum cost, and therefore terminate calculation.

(3) Increase flow. The same procedures as finding maximum flow, i.e., let

$$d_{ij} = c_{ij} - f_{ij}, \quad \text{if } e_{ij} \in h^+;$$

$$d_{ij} = f_{ij}, \quad \text{if } e_{ij} \in h^-;$$

$$d = \min\{d_{ij}|e_{ij} \in h\}.$$

Define the flow $f = \{fij\}$ as

$$fij = fij + d, \quad \text{if } e_{ij} \in h^+;$$

$$fij = fij - d, \quad \text{if } e_{ij} \in h^-;$$

$$fij = fij, \quad \text{others.}$$

If M_f is greater than or equal to the desired flux, then reasonably reduce d, such that M_f is equal to the desired flux, then f is the minimum cost flow and terminate calculation; or else return to (1).

The Java codes, minCost, for minimum cost flow algorithm are:

```
//Minimum cost flow algorithm.
/*v: number of vertice; Data are stored in Two Array Listing.
d1[1-e], d2[1 - e]: start and end vertice of edges; cc[1 - e]:
flow capacity of edges; bb[1 - e]: cost of unit flow of edges;
Vertice are sequentially numbered in the graph. */
    public class minCost {
    public static void main(String[] args){
    int i,e;
    if (args.length!=1)
    System.out.println("You must input the name of table in the
database. For example, you may type the following in the command
window: java minCost mincost, where mincost is the name of
table.");
    String tablename = args[0];
    readDatabase readdata = new readDatabase("dataBase",
tablename,4);
    e = readdata.m;
    int a[] = new int[e+1];
    int b[] = new int[e+1];
    double c[] = new double[e+1];
    double d[] = new double[e+1];
```

```
    for(i=1;i<=e;i++) {
    a[i] = (Integer.valueOf(readdata.data[i][1])).intValue();
    b[i] = (Integer.valueOf(readdata.data[i][2])).intValue();
    c[i] = (Double.valueOf(readdata.data[i][3])).
doubleValue();
    d[i] = (Double.valueOf(readdata.data[i][4])).
doubleValue(); }
    minCost(e,a,b,c,d); }
    public static void minCost(int e, int d1[], int d2[],
double cc[], double bb[]) {
    int i, j, k, v, m, d;
    double mf,mmf,mf0,dv,dvt,inf = 1e+30;
    v = 0;
    for(i=1;i<=e;i++) {
    if (d1[i]>v) v = d1[i];
    if (d2[i]>v) v = d2[i]; }
    double c[][] = new double[v+1][v+1];
    double f[][] = new double[v+1][v+1];
    double a[][] = new double[v+1][v+1];
    double b[][] = new double[v+1][v+1];
    double p[] = new double[v+1];
    int s[] = new int[v+1];
    for(i=1;i<=v;i++)
    for(j=1;j<=v;j++)
    for(k=1;k<=e;k++)
    if ((d1[k]==i) & (d2[k]==j)) {
    c[i][j] = cc[k];
    b[i][j] = bb[k];
    break; }
    mf = 0;
    mf0 = inf;
    for(i=1;i<=v;i++)
    for(j=1;j<=v;j++) f[i][j]=0;
    while(v>0) {
    for(i=1;i<=v;i++)
    for(j=1;j<=v;j++)
    if (j!=i) a[i][j]=inf;
    for(i=1;i<=v;i++)
    for(j=1;j<=v;j++)
    if ((c[i][j]>0) & (f[i][j]==0)) a[i][j]=b[i][j];
    else if ((c[i][j]>0) & (f[i][j]==c[i][j]))
a[j][i]=-b[i][j];
    else if (c[i][j]>0) {
    a[i][j] = b[i][j];
    a[j][i] = - b[i][j]; }
    for(i=2;i<=v;i++) {
```

```
p[i] = inf;
s[i] = i; }
for(k=1;k<=v;k++) {
d = 1;
for(i=2;i<=v;i++)
for(j=1;j<=v;j++)
if (p[i]>(p[j]+a[j][i])) {
p[i] = p[j]+a[j][i];
s[i] = j;
d = 0; }
if (d!=0) break; }
if (p[v]>=inf) break;
dv = inf;
m = v;
while(v>0) {
dvt =  -inf;
if (a[s[m]][m]>0) dvt = c[s[m]][m] - f[s[m]][m];
else if (a[s[m]][m]<0) dvt = f[m][s[m]];
if (dv>dvt) dv = dvt;
if (s[m]==1) break;
m = s[m]; }
d = 0;
if ((mf+dv)>=mf0) {
dv = mf0 - mf;
d = 1; }
m = v;
while (v>0) {
if (a[s[m]][m]>0) f[s[m]][m]+=dv;
else if (a[s[m]][m]<0) f[m][s[m]]-=dv;
if (s[m]==1) break;
m = s[m]; }
if (d!=0) break;
mf = 0;
for(j=1;j<=v;j++) mf+=f[1][j]; }
mmf = 0;
for(i=1;i<=v;i++)
for(j=1;j<=v;j++) mmf+=b[i][j]*f[i][j];
System.out.println("Maximum flow with minimum cost:");
for(i=1;i<=v;i++) {
for(j=1;j<=v;j++)
System.out.print(f[i][j]+" ");
System.out.println(); }
System.out.println("Maximum flow with minimum cost="+mf);
System.out.println("Minimum cost="+mmf); }
}
```

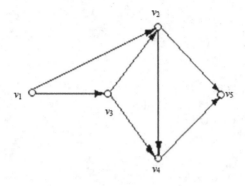

Figure 4.   A ecotour network.

For example, in the ecotour network of Fig. 4, there are five scenic spots and seven paths. Human crowd flux and cost per unit flux of every path are indicated in Fig. 4. The data of human crowd flux and cost per unit flux are listed as follows:

$$d_1 = (v_1, v_1, v_2, v_2, v_3, v_3, v_4),$$

$$d_2 = (v_2, v_3, v_4, v_5, v_2, v_4, v_5),$$

$$cc = (7, 8, 5, 6, 9, 8, 6).$$

$$bb = (3, 2, 7, 3, 2, 2, 1).$$

According to the minimum cost flow algorithm, the minimum cost and maximum human crowd flow matrix is obtained.

$$\begin{bmatrix} 0 & 6 & 6 & 0 & 0 \\ 0 & 0 & 0 & 0 & 6 \\ 0 & 0 & 0 & 6 & 0 \\ 0 & 0 & 0 & 0 & 6 \\ 0 & 0 & 0 & 0 & 0 \end{bmatrix}$$

The minimum cost is 66, maximum human crowd flux with minimum cost is 12.

2.4. Shortest tree

Suppose a graph X has v vertices and e edges. Sum of weights of all tree branches of a tree is called weight of the tree (tree weight). A tree with minimal tree weight is called the shortest tree (minimum spanning tree). A method to find the shortest tree is (Lu and Lu, 1995):

(1) Arbitrarily use a spanning tree.
(2) Add an edge of a cotree to form a circuit. In the circuit, if there is an edge which is longer than the edge added, then replace the longer edge with the new added edge and thus achieve a new tree. Repeat this process until longer edge is no more present.

To find the shortest tree, Kruskal algorithm is always used. In the algorithm, check the edges of X from smaller weight edge to larger weight edge, add them to T based on the principle of not generating any loop, until the number of edges of T is equal to (the number of vertices of $X - 1$). The procedures are as follows:

(1) Order the edges in the edge set, from smaller weight edge to larger weight edge, as $e_1, e_2, \ldots, e_v$.
(2) Let $T = \{e_1\}, i = 1, j = 2$.
(3) If $i = v - 1$, then obtain the output of T and terminate calculation, or else return to (4).
(4) If a circuit is generated after e_i is added to T, then let $j = j + 1$, return to (4), or else return to (5).
(5) Let $T = T \cup \{e_i\}, j = j + 1, i = i + 1$, return to (3).

The following are Java codes, Kruskal, for Kruskal algorithm:

```
/*Kruskal algorithm to calculate the shortest tree in an
undirected graph.*/
/*v: number of vertice; a[1 - v][1 - v]: between-vertex
weights */
    public class Kruskal {
    public static void main(String[] args){
    int i, j, v;
    if (args.length!=2)
    System.out.println("You must input the number of samplings,
and the name of table in the database. For example, you may type
the following in the command window: java Kruskal 5 kruskal,
where kruskal is the name of table, 5 is the number of vertice.");
    v = Integer.valueOf(args[0]).intValue();
    String tablename = args[1];
    readDatabase                          readdata = new
    readDatabase("dataBase",tablename,v);
```

```
double a[][] = new double[v+1][v+1];
for(i=1;i<=v;i++)
for(j=1;j<=v;j++)
a[i][j] = (Double.valueOf(readdata.data[i][j])).
doubleValue();
Kruskal(v, a); }
public static void Kruskal(int v, double a[][]) {
int i,j,k,l,kk,i1,j1,m,in,c;
double cc = 0;
double t[][] = new double[v+1][v+1];
double t1[][] = new double[v+1][v+1];
double b[] = new double[10000];
k = 1;
for(i=1;i<=v-1;i++)
for(j=i+1;j<=v;j++)
if (a[i][j]>0) {
b[k] = a[i][j];
kk = 1;
for(l=1;l<=k-1;l++)
if (b[k]==b[l]) {
kk=0;
break; }
k+=kk; }
k--;
for(i=1;i<=k-1;i++)
for(j=i+1;j<=k;j++)
if (b[j]<b[i]) {
cc = b[j];
b[j] = b[i];
b[i] = cc; }
m = 0;
for(l=1;l<=k;l++) {
if (m==v) break;
for(i=1;i<=v-1;i++)
for(j=i+1;j<=v;j++)
if (a[i][j]==b[l]) {
t[i][j] = b[l];
t[j][i] = b[l];
for(i1=1;i1<=v;i1++)
for(j1=1;j1<=v;j1++)
t1[i1][j1]=t[i1][j1];
while(v>0) {
in = 1;
c = 0;
for(i1=1;i1<=v;i1++) {
kk = 0;
```

```
for(j1=1;j1<=v;j1++)
if (t1[i1][j1]>0) {
kk++;
c = j1; }
if (kk==1) {
t1[i1][c] = 0;
t1[c][i1] = 0;
in = 0; } }
if (in!=0) break; }
in = 0;
for(i1=1;i1<=v-1;i1++)
for(j1=i1+1;j1<=v;j1++)
if (t1[i1][j1]>0) {
in = 1;
break; }
if (in!=0) {
t[i][j]=0;
t[j][i]=0; }
else m++; } }
System.out.println("Shortest tree:");
for(i=1;i<=v;i++) {
for(j=1;j<=v;j++)
System.out.print(t[i][j]+" ");
System.out.println(); } }
}
```

For example, for the cost per unit flux in Fig. 4, its weight matrix is:

$$\begin{bmatrix} 0 & 3 & 2 & 0 & 0 \\ 0 & 0 & 0 & 7 & 3 \\ 0 & 2 & 0 & 2 & 0 \\ 0 & 0 & 0 & 0 & 1 \\ 0 & 0 & 0 & 0 & 0 \end{bmatrix}$$

According to Kruskal algorithm, the shortest tree is:

$$\begin{bmatrix} 0 & 3 & 2 & 0 & 0 \\ 3 & 0 & 0 & 0 & 0 \\ 2 & 0 & 0 & 2 & 0 \\ 0 & 0 & 2 & 0 & 1 \\ 0 & 0 & 0 & 1 & 0 \end{bmatrix}$$

i.e., it is (v_2, v_1), (v_1, v_3), (v_3, v_4), (v_4, v_5).

2.5. *Matching problem*

Matching problem is an important part of graph theory. It has important applications in the optimal assignment problem.

Let M be a subset of the edge set E of a graph X. If any two edges of M are disjoint in X, then M is called a matching of X. The two vertices of an edge of M are called matched in M. If an edge of the matching M is associated with the vertex v, then M saturates vertex v, otherwise v is called M-unsaturated. If every vertex of X is M-saturated, then the matching M is called the optimum matching. If there is no other matching M' in X, such that $|M'| > |M|$, then M is called a maximum matching of X. Every optimum matching is a maximum matching.

In a graph X, the M-augmenting path refers to a taggered path that both initial vertex and terminal vertex are M-unsaturated.

Theorem 4. *The matching M of graph X is a maximum matching if and only if there is no M-augmenting path in X.*

2.5.1. *Optimum matching*

The procedures for optimum matching are as follows.

Suppose the graph $X = (U, V, C, D)$ is a complete bipartite and weighted graph, L is a label for initial available vertex. Let

$$L(p) = \max\{D(pq)|q \in V\}, \quad p \in U;$$

$$L(q) = 0, \quad q \in V.$$

M is a matching of X_L.

(1) If every vertex of U is M-saturated, then M is the optimum matching, otherwise find a M-unsaturated vertex $w \in U$, let $S = \{w\}$, $T = \phi$, return to (2).
(2) Let $N_L(S) = \{v|w \in S, wv \in X_L\}$. If $N_L(S) = T$, then X_L has no optimum matching, otherwise return to (3); or else return to (4).
(3) Adjust the label of available vertex. Calculate

$$I_L = \min\{L(p) + L(q) - D(pq)|p \in S, q \in V - T\}$$

to obtain the new available vertex label:

$$G(v) = L(v) - I_L, \quad v \in S;$$

$$G(v) = L(v) + I_L, \quad v \in T;$$

$$G(v) = L(v), \quad \text{others}.$$

Let $L = G$, $X_L = X_G$, once again give a matching M of X_L, return to(1).

(4) Choose $q \in N_L(S) - T$, if q is M-saturated, return to (5), otherwise return to (6).

(5) If $pq \in M$, then let $S = S \cup \{p\}$, $T = T \cup \{q\}$, return to (2).

(6) The w–q path in X_L is the M-augmenting path, denoted by P, and let $M = M \oplus P$, return to (1), where $M \oplus P = M \cup P - M \cap P$.

The following are Java codes (to be revised), optMatch, for optimum matching algorithm:

```
//Optimum matching algorithm.
/*v: number of vertice; e: number of edges; Data are stored
in Two Array Listing. d1[1-e], d2[1-e]: start and end vertice
of edges; dd[1-e]: cost of edges; Vertice are sequentially
numbered in the graph. */
public class optMatch {
public optMatch(int d1[], int d2[], double dd[]) {
private int i, j, k, i1, j1, v, e, kr, pn, ps, pt;
private double max, il;
e = d1.length;
v = 0;
for(i=1;i<=e;i++) {
if (d1[i]>v) v=d1[i];
if (d2[i]>v) v=d2[i]; }
private double l[][]=new double[v+1][v+1];
private double ia[][]=new double[v+1][v+1];
private double xl[][]=new double[v+1][v+1];
private int m[][]=new int[v+1][v+1];
private int s[]=new int[v+1];
private int t[]=new int[v+1];
private int nl[]=new int[v+1];
for(i=1;i<=v;i++)
for(j=1;j<=v;j++)
for(k=1;k<=e;k++)
if ((d1[k]==i) & (d2[k]==j)) {
```

```
ia[i][j] = dd[k];
break; }
for(i=1;i<=v;i++) {
l[i][1] = 0;
l[i][2] = 0; }
for(i=1;i<=v;i++)
for(j=1;j<=v;j++) {
if (l[i][1]<ia[i][j]) l[i][1] = ia[i][j];
m[i][j] = 0; }
for(i=1;i<=v;i++)
for(j=1;j<=v;j++)
if ((l[i][1]+l[j][2])==ia[i][j]) xl[i][j]=1;
else xl[i][j] = 0;
i1 = 0;
j1 = 0;
for(i=1;i<=v;i++) {
for(j=1;j<=v;j++)
if (xl[i][j]!=0) {
i1 = i;
j1 = j;
break; }
if (i1!=0) break; }
m[i1][j1]=1;
for(i=1;i<=v;i++) {
s[i] = 0;
t[i] = 0;
nl[i] = 0; }
while (v>0) {
for(i=1;i<=v;i++) {
k = 1;
for(j=1;j<=v;j++)
if (m[i][j]!=0) {
k = 0;
break; }
if (k!=0) break; }
if (k==0) break;
s[1] = i;
ps = 1;
pt = 0;
while (v>0) {
pn = 0;
for(i=1;i<=ps;j++)
for(j=1;j<=v;j++)
if (xl[s[i]][j]!=0) {
pn++;
nl[pn]=j;
```

```
        for(k=1;k<=pn-1;k++)
        if (nl[k]==j) pn--; }
        if (pn==pt) {
        kr = 1;
        for(j=1;j<=pn;j++)
        if (nl[j]!=t[j]) {
        kr = 0;
        break; } }
        if ((pn==pt) & (kr!=0)) {
        il = inf;
        for(i=1;i<=ps;i++)
        for(j=1;j<=n;j++) {
        kr = 1;
        for(k=1;k<=pt;k++)
        if (t[k]==j) {
        kr = 0;
        break; }
        if ((kr!=0) & (il>(l[s[i]][1]+l[j][2]-ia[s[i]][j]))))
il=l[s[i]][1]+l[j][2]-ia[s[i]][j]; }
        for(i=1;i<=ps;i++) l[s[i]][1]-=il;
        for(j=1;j<<=pt;j++) l[t[j]][2]+=il;
        for(i=1;i<=v;i++)
        for(j=1;j<=v;j++) {
        if ((l[i][1]+l[j][2])==ia[i][j]) xl[i][j]=1;
        else xl[i][j] = 0;
        m[i][j] = 0;
        k = 0; }
        i1 = 0;
        j1 = 0;
        for(i=1;i<=v;i++) {
        for(j=1;j<=v;j++)
        if (xl[i][j]!=0) {
        i1 = i;
        j1 = j;
        break; }
        if (i1!=0) break; }
        m[i1][j1]=1;
        break; }
        else {
        for(j=1;j<=pn;j++) {
        kr = 1;
        for(k=1;k<=pt;k++)
        if (t[k]==nl[j]) {
        kr = 0;
        break; }
        if (kr!=0) {
```

```
j1 = j;
break; } }
kr = 0;
for(i=1;j<=v;j++)
if (m[i][nl[j1]]!=0) {
kr = 1;
i1 = i;
break; }
if (kr!=0) {
ps++;
s[ps]=i1;
pt++; }
t[pt]=nl[j1];
else {
for(k=1;k<=pt;k++) {
m[s[k]][t[k]]=1;
m[s[k+1]][t[k]]=0; }
if (pt==0) k=0;
m[s[k+1]][nl[j1]]=1;
break; } } } }
max=0;
for(i=1;i<=v;i++)
for(j=1;j<=v;j++)
if (m[i][j]!=0) max+=ia[i][j];
System.out.println("Optimum matching:");
for(i=1;i<=v;i++) {
for(j=1;j<=v;j++)
System.out.print(m[i][j]+" ");
System.out.println(); }
System.out.println("Weight for optimum matching=:
"+max); }
}
```

For example, given that we want to green four hills I, II, III, and IV, using four tree species, A, B, C, and D. The costs are listed below. Each tree species must only green one hill. We want to know which plan will cost least.

$$d_1 = (v_1, v_1, v_1, v_1, v_2, v_2, v_2, v_2, v_3, v_3, v_3, v_3, v_4, v_4, v_4),$$

$$d_2 = (v_1, v_2, v_3, v_4, v_1, v_2, v_3, v_4, v_1, v_2, v_3, v_4, v_1, v_3, v_4),$$

$$dd = (3, 4, 2, 6, 2, 7, 1, 5, 2, 6, 1, 4, 8, 1, 3).$$

Using the optimum matching algorithm, we obtain the optimum matching matrix:

$$\begin{bmatrix} 0 & 0 & 0 & 1 \\ 0 & 0 & 1 & 0 \\ 0 & 1 & 0 & 0 \\ 1 & 0 & 0 & 0 \end{bmatrix}$$

i.e., the plan is: (A, IV), (B, III), (C, II), (D, I). The weight of the optimum matching matrix is 21.

2.5.2. *Maximum matching*

Here we discuss an algorithm to find maximum matching of a bipartite graph, i.e., Hungarian algorithm.

An example for maximum matching problem of a bipartite graph is that suppose there are m tree species with which to green n hills, we want to find a plan that optimizes biodiversity conservation. Here any of the m tree species can be planted on one or more hills. However, not all tree species will surely be planted on any hill. The problem is whether we can assign each tree species to a hill for planting.

Here we denote m tree species and n hills as $G = \{g_1, g_2, \ldots, g_m\}$ and $H = \{h_1, h_2, \ldots, h_n\}$, respectively. g_i and h_j are adjacent if and only if the tree species g_i can be planted on the hill h_j. The problem is thus to find a maximum matching of graph X.

The principle of Hungarian algorithm is, starting from any matching M of a graph X, search for M-augmenting path for all M-unsaturated vertices in X. If there is no M-augmenting path, then M is the maximum matching. If there exists a M-augmenting path C, then exchange the M edges and non-M edges in C to obtain a matching M_1 which has one more edge than M. Repeat the above process on M_1. Suppose $X = (G, H, E)$ is a bipartite graph, where $G = \{g_1, g_2, \ldots, g_m\}$, $H = \{h_1, h_2, \ldots, h_n\}$. Arbitrarily find an initial matching M of X, and then

(1) Let $S = \phi$, $T = \phi$, return to (2).
(2) If M saturates all vertices of $G - S$, then M is the maximum matching of bipartite graph X. Otherwise, arbitrarily find a M-unsaturated vertex $u \in G - S$, let $S = S \cup \{u\}$, return to (3).

(3) Let $N(S) = \{v | u \in S, uv \in E\}$. If $N(S) = T$, return to (2). Otherwise, find $h \in N(S) - T$. If h is M-saturated, return to (4), otherwise return to (5).

(4) Suppose $gh \in M$, then let $S = S \cup \{g\}$, $T = T \cup \{h\}$, return to (3).

(5) $u - h$ path is a M-augmenting path, denoted by C, and let $M = M \oplus C$, return to (1), where, $M \oplus C = M \cup C - M \cap C$.

The Java codes, maxMatch, for maximum matching algorithm that uses labels are:

```
/*Maximum matching algorithm. v: number of elements of G;
n: number of elements of H; a[1 - v][1 - n]: data matrix. */
public class maxMatch {
public static void main(String[] args){
int i,j,n,v;
if (args.length!=2)
System.out.println("You must input the number of samplings,
and the name of table in the database. For example, you may type
the following in the command window: java maxMatch 5 maxmatch,
where maxmatch is the name of table, 5 is the number of columns
in the table.");
n = Integer.valueOf(args[0]).intValue();
String tablename = args[1];
readDatabase readdata = new readDatabase("dataBase",
tablename,n);
v = readdata.m;
double a[][]=new double[v+1][n+1];
for(i=1;i<=v;i++)
for(j=1;j<=n;j++)
a[i][j] = (Double.valueOf(readdata.data[i][j])).
doubleValue();
maxMatch(v,n,a); }
public static void maxMatch(int v, int n, double a[][]) {
int i,j,k,cd,cp=0,gi;
int c[][] = new int[n+1][5];
int m[][] = new int[v+1][n+1];
int g[] = new int[v+1];
int h[] = new int[n+1];
int hh[] = new int[n+1];
for(i=1;i<=v;i++) {
for(j=1;j<=n;j++)
if (a[i][j]!=0) {
m[i][j]=1;
break; }
```

```
if (m[i][j]!=0) break; }
while(v>0) {
for(i=1;i<=v;i++) g[i]=0;
for(i=1;i<=n;i++) h[i]=0;
for(i=1;i<=v;i++) {
cd = 1;
for(j=1;j<=n;j++)
if (m[i][j]!=0) cd=0;
if (cd!=0) g[i]=-n-1; }
cd = 0;
while(v>0) {
gi = 0;
for(i=1;i<=v;i++)
if (g[i]<0) {
gi = i;
break; }
if (gi==0) {
cd = 1;
break; }
g[gi]*=(-1);
k = 1;
for(j=1;j<=n;j++)
if ((a[gi][j]!=0) & (h[j]==0)) {
h[j] = gi;
hh[k] = j;
k++; }
if (k>1) {
k--;
for(j=1;j<=k;j++) {
cp = 1;
for(i=1;i<=v;i++)
if (m[i][hh[j]]!=0) {
g[i] = —hh[j];
cp = 0;
break; }
if (cp!=0) break; }
if (cp!=0) {
k = 1;
j = hh[j];
while(v>0) {
c[k][2] = j;
c[k][1] = h[j];
j = Math.abs(g[h[j]]);
if (j==(n+1)) break;
k++; }
for(i=1;i<=k;i++)
```

```
if (m[c[i][1]][c[i][2]]!=0) m[c[i][1]][c[i][2]]=0;
else m[c[i][1]][c[i][2]]=1;
break; } } }
if (cd!=0) break; }
System.out.println("Maximum matching:");
for(i=1;i<=v;i++) {
for(j=1;j<=n;j++)
System.out.print(m[i][j]+" ");
System.out.println(); } }
}
```

Suppose there are six tree species which are used to green five hills (Fig. 5). We want to find a plan so that the biodiversity conservation is maximized. The data for Fig. 5 is provided in the matrix:

$$\begin{bmatrix} 1 & 1 & 1 & 0 & 0 \\ 1 & 1 & 0 & 1 & 0 \\ 0 & 1 & 1 & 0 & 1 \\ 0 & 1 & 1 & 0 & 0 \\ 0 & 1 & 0 & 1 & 1 \\ 0 & 0 & 0 & 1 & 1 \end{bmatrix}$$

The maximum matching matrix of the Hungarian algorithm is:

$$\begin{bmatrix} 1 & 0 & 0 & 0 & 0 \\ 0 & 1 & 0 & 1 & 0 \\ 0 & 0 & 1 & 0 & 0 \\ 0 & 1 & 0 & 0 & 0 \\ 0 & 0 & 0 & 0 & 1 \\ 0 & 0 & 0 & 0 & 0 \end{bmatrix}$$

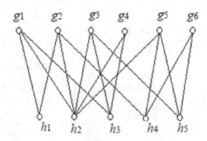

Figure 5.   A bipartite graph for hills' greening.

Therefore, the maximum matching is: (tree 1, hill 1), (tree 2, hill 2, hill 4), (tree 3, hill 3), (tree 4, hill 2), (tree 5, hill 5).

3. Determination of Network Weights

3.1. *Analytic hierarchy process (AHP)*

Analytic Hierarchy Process (AHP) can be used in network analysis for determining weights of edges and vertices of a network.

AHP is suitable for the hierarchical structure problems that are hard to be quantified, as indicated in Fig. 6.

In AHP of network analysis, we need to organize all vertices into several layers according to their attributes and relations. The top layer contains one vertex. Between-layer vertices are linked by directed edges.

In general, the number of layers is infinite. However, too many adjacent vertices or edges will result in a difficulty in pairwise comparisons.

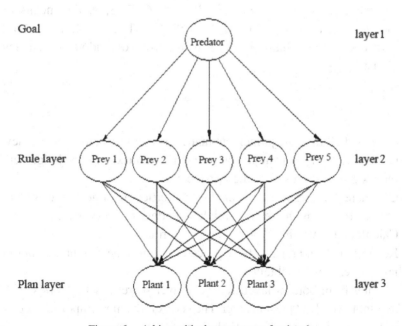

Figure 6.   A hierarchical structure on food web.

In the network analysis, AHP can be conducted in the following way:

(1) **Establish a hierarchical structure model.** Determine the adjacent vertices at lower layer for every vertex. Starting from the top layer, for every layer, label all vertices from left to right at each layer.

(2) **Construct judgement matrix of every layer.** A judgement matrix is a weight matrix in which an element is the weight of a vertex over its adjacent vertices (or its associated edges) at lower layer. The lower triangular vertices in a judgement matrix are the reciprocals of corresponding vertices in upper triangular matrix, and the diagonal elements are 1's.

(3) **Consistency test of judge matrix.** For a judgement matrix, $B = (b_{ij})_{m \times m}$, calculate maximum eigenvalue $\lambda_{\max}$, and the eigenvector.

Consistency CI of a judgement matrix B is

$$CI = (\lambda_{\max} - m)(m - 1).$$

A smaller CI means that the judgement matrix is close to complete consistency. When $m = 1, 2, 3, 4, 5, 6, 7, 8, 9$, the means of random consistency RI are 0, 0, 0.58, 0.9, 1.12, 1.24, 1.32, 1.41, 1.45, respectively (Jiang, 1988). The proportion of random consistency CR is:

$$CR = CI/RI.$$

If $CR < 0.1$, then the judgement matrix possesses a better consistency. Otherwise, we need to adjust the judgement matrix, until the judgement matrix possesses a satisfied consistency.

(4) Calculate the weights of present layer's vertices over next layer's adjacent vertices, which are also weights of edges of every vertex.

(5) Calculate the weights of lower layer's vertices.

(6) Return to (3), and repeat above procedures, until the weights of bottom layer's vertices are obtained.

(7) Starting from bottom layer, for each layer search for the vertex with maximal weight at present layer. The vertex is the most important vertex of the present layer relative to the top layer vertex.

The Java codes, AHP, for AHP algorithm are as follows:

```
/*AHP algorithm for multi-phase problems.*/
  public class AHP {
  public static void main(String[] args){
  if (args.length!=1) System.out.println("You must input the
name of table in the database. For example, you may type the
following in the command window: java AHP ahp, where ahp is
the name of table.");
  String tablename=args[0];
  readDatabase readdata=new readDatabase("dataBase",
tablename, 5);
  int m,ii,jj,k,rr,mm;
  double opt;
  mm=readdata.m;
  int d1[]=new int[mm+1];
  int d11[]=new int[mm+1];
  int d2[]=new int[mm+1];
  int d22[]=new int[mm+1];
  int d3[]=new int[mm+1];
  int d33[]=new int[mm+1];
  int d4[]=new int[mm+1];
  int d44[]=new int[mm+1];
  int u[]=new int[mm+1];
  int lab[]=new int[mm+1];
  double d[]=new double[mm+1];
  double dd[]=new double[mm+1];
  double ww[][]=new double[4][mm+1];
  double w[]=new double[mm+1];
  double wa[]=new double[mm+1];
  double wb[]=new double[mm+1];
  for (int i=1;i<=mm;i++) {
  d1[i]=(Integer.valueOf(readdata.data[i][1])).intValue();
  d2[i]=(Integer.valueOf(readdata.data[i][2])).intValue();
  d3[i]=(Integer.valueOf(readdata.data[i][3])).intValue();
  d4[i]=(Integer.valueOf(readdata.data[i][4])).intValue();
  d[i]=(Double.valueOf(readdata.data[i][5])).
doubleValue(); }
  k=ii=rr=0;
  while(ii<mm) {
  int nn=0;
  for (int j=k+1;j<=mm;j++)
  if ((d1[j]==d1[k+1]) & (d2[j]==d2[k+1])) nn++;
  else break;
  for (int i=k+1;i<=k+nn;i++) {
  d11[i − k]=d1[i];
```

```
d22[i − k]=d2[i];
d33[i − k]=d3[i];
d44[i − k]=d4[i];
dd[i − k]=d[i]; }
m = (int)(1+Math.sqrt(1+8*nn))/2;
weight(m,d11,d22,d33,d44,dd,w,u);
rr+=m;
for(int i=rr-m+1;i<=rr;i++) {
ww[1][i] = d11[i - rr+m];
ww[2][i] = u[i - rr+m];
ww[3][i]=w[i - rr+m]; }
k+ = nn;
ii+ = nn; }
int la = 0;
for(int i=1;i<=rr;i++)
if (ww[1][i]>la) la=(int)ww[1][i];
int ls = 0;
for(int i=1;i<=rr;i++)
if (((int)ww[1][i]==la) & (ww[2][i]>ls)) ls=(int)ww[2][i];
for(int i=1;i<=la;i++) {
ii = 0;
k = 0;
for (int j=1;j<=rr;j++) {
if ((int)ww[1][j]==i) k++;
if (((int)ww[1][j]==i) & (ww[2][j]>ii)) ii++; }
d[i] = ii;
u[i] = k; }
u[la+1] = ls;
d[0] = 1;
d[la+1] = ls;
System.out.println("Number of vertices for every layer:");
for(int i=0;i<=la;i++)
System.out.print("Layer-"+(i+1)+"="+(int)d[i]+" ");
System.out.println("\n");
System.out.println("Weight of layer-1 vertex:\n1.0\n");
System.out.println("Weights of layer-2 vertices:");
opt = 0;
d11[2] = 1;
for(int i=1;i<=u[1];i++) {
wa[i] = ww[3][i];
System.out.print(wa[i]+" ");
if (wa[i]>opt) {
opt = wa[i];
d11[2] = i; } }
System.out.println("\n");
double ss;
```

```
for(int i = 2;i<=la;i++) {
ss=jj=0;
for (int j=u[i-1]+1;j<=u[i-1]+u[i];j++) {
ss+=ww[3][j];
if (Math.abs(ss-1)<1e-06) {
jj++;
lab[jj]=j;
ss = 0; } }
for(k=1;k<=d[i];k++) {
ss = 0;
for (int j=u[i-1]+1;j<=u[i-1]+u[i];j++)
if ((int)ww[2][j]==k) {
for (ii=0;ii<=jj-1;ii++)
if ((j<=lab[ii+1]) & (j>lab[ii])) break;
ss+=wa[ii+1]*ww[3][j]; }
wb[k] = ss; }
System.out.println("Weights of layer-"+(i+1)+"
vertices:");
opt = 0;
d11[i+1]=1;
for(k=1;k<=d[i];k++) {
wa[k] = wb[k];
System.out.print(wa[k]+" ");
if (wa[k]>opt) {
opt = wa[k];
d11[i+1] = k; } }
System.out.println("\n"); }
System.out.println("Most important vertex in every layer in
respect to the start vertex::");
for(k=2;k<=la+1;k++)
System.out.print("Layer-"+k+"="+d11[k]+" ");
}
public static double[] weight(int m, int d11[],
int d22[], int d33[], int d44[], double dd[], double w[],
int u[]) {
double[] ri={0,0.00,0.00,0.58,0.90,1.12,1.21,1.32,1.41,
1.45,1.49};
double ci,cr,lamda,s;
double w1[]=new double[m+1];
double b[][]=new double[m+1][m+1];
int vv=0;
for(int i=1;i<=m;i++)
for(int j=i;j<=m;j++) {
if (j==i) {
b[i][j]=1;
continue; }
```

```
if (j!=i) {
vv++;
b[i][j] = dd[vv]; }
b[j][i] = 1/b[i][j]; }
s = 0;
for(int i=1;i<=m;i++) {
w[i] = b[i][1];
for(int j=2;j<=m;j++)
w[i] = w[i]*b[i][j];
w[i] = Math.pow(w[i],1.0/m);
s+=w[i]; }
for(int i=1;i<=m;i++)
w[i] = w[i]/s;
for(int i=1;i<=m;i++) {
w1[i] = 0;
for(int j=1;j<=m;j++)
w1[i]+=b[i][j]*w[j]; }
lamda = 0;
for(int i=1;i<=m;i++)
lamda+=w1[i]/w[i];
lamda/=m;
ci = (lamda-m)/(m-1);
cr = ci/ri[m];
int ii = 1,jj;
u[ii] = d33[1];
for(int k=1;k<=(m*m-m)/2;k++) {
vv = ii;
jj = 0;
for(int i=1;i<=vv;i++)
if (u[i]!=d33[k]) jj++;
if (jj==vv) {
ii++;
u[ii] = d33[k]; }
vv = ii;
jj = 0;
for(int i=1;i<=vv;i++)
if (u[i]!=d44[k]) jj++;
if (jj==vv) {
ii++;
u[ii] = d44[k]; } }
System.out.println("Consistency CI and random consistencey
ratio CR of the matrix of next-layer vertices to the vertex
"+d22[1]+" of layer "+d11[1]+":");
System.out.println("CI="+ci+" "+"CR="+cr);
if (cr<0.1) {
System.out.println("The matrix is consistent. It is OK.");
```

```
System.out.println("Weights of next-layer vertices to the
vertex "+d22[1]+" of layer "+d11[1]+":");
    for(int i=1;i<=m;i++)
    System.out.println("Weight of next-layer vertex
"+u[i]+"="+w[i]+" ");
    System.out.println(); }
    else {
    System.out.println("Due to CR>0.1, matrix of next-layer
vertices to the vertex "+d22[1]+" of layer "+d11[1]+ " is not
consistent. Please try to adjust this matrix and run one more
time.");
    System.exit(0); }
    return w; }
    }
```

For example, a food web has three trophic layers. Layer 1 is a predator, layer 2 has five prey species being predated by the predator, and layer 3 has three plant species being fed by preys. We want to determine the most important plan species for the predator.

Five prey species were compared against their importance relative to the predator. The resulting judgement matrix is:

$$
\begin{bmatrix}
1 & 3 & 2 & 3 & 4 \\
1/3 & 1 & 3 & 1 & 3 \\
1/2 & 1/3 & 1 & 1/3 & 3 \\
1/3 & 1 & 3 & 1 & 6 \\
1/4 & 1/3 & 1/3 & 1/6 & 1
\end{bmatrix}
$$

Three plant species were compared against their importance relative to prey species 1, and the judgement matrix is:

$$
\begin{bmatrix}
1 & 1 & 2 \\
1 & 1 & 5 \\
1/2 & 1/5 & 1
\end{bmatrix}
$$

In the same way, we obtained the remaining four judgement matrices for other four prey species.

Two Array Listing is used to record the hierarchical structures and judgement matrices of AHP. For example, the graph in Fig. 6 is recorded in Two Array Listing as the following:

$$d_1 = (1, 1, 1, 1, 1, 1, 1, 1, 1, 1, 2, 2, 2, 2, 2, 2, 2, 2, 2, 2, 2, 2, 2, 2, 2),$$

$$d_2 = (1, 1, 1, 1, 1, 1, 1, 1, 1, 1, 1, 1, 1, 2, 2, 2, 3, 3, 3, 4, 4, 4, 5, 5, 5),$$

$$d_3 = (1, 1, 1, 1, 2, 2, 2, 3, 3, 4, 1, 1, 2, 1, 1, 2, 1, 1, 2, 1, 1, 2, 1, 1, 2),$$

$$d_4 = (2, 3, 4, 5, 3, 4, 5, 4, 5, 5, 2, 3, 3, 2, 3, 3, 2, 3, 3, 2, 3, 3, 2, 3, 3),$$

$$d = (3, 5, 3, 4, 3, 1, 3, 1/3, 3, 6, 1, 2, 5, 3, 8, 5, 2, 5, 2, 1/2, 1/3,$$
$$1, 1, 1/4, 2),$$

where d_1, d_2, d_3, d_4, are labels of present layers, labels of present layer's vertices (from left to right), labels of next layer's vertices (from left to right), and labels of next layer's vertices (from left to right), and d is the weight of every vertex at next layer over its followed vertex at the same layer.

Obviously, we need only to store the upper triangular elements of the judgement matrix. Diagonal elements are 1's.

According to calculation, all judgement matrices possess satisfied consistency.

At layer 2, the weights of prey 1 to 5 are 0.388, 0.205, 0.115, 0.236, 0.056, respectively.

The weights of three edges of prey 1 at layer 2 are 0.367, 0.498, 0.135, respectively.

The weights of three edges of prey 2 at layer 2 are 0.661, 0.272, 0.067, respectively.

The weights of three edges of prey 3 at layer 2 are 0.595, 0.276, 0.128, respectively.

The weights of three edges of prey 4 at layer 2 are 0.169, 0.387, 0.444, respectively.

The weights of three edges of prey 5 at layer 2 are 0.174, 0.192, 0.634, respectively.

Three weights of plant species 1 to 3 at layer 3 are 0.396, 0.383, 0.221, respectively.

It is obvious that plant species 1 is the most important to the predator, followed by plant species 2. Among all prey species, prey species 1 is the most important to the predator.

3.2. *Neural network algorithms*

By knowing the connection structure and input-output data of a network, various neural network algorithms (Zhang, 2010) can be revised and used to determine the weights of network edges. A neural network algorithm can adjust and determine the weights of network edges by learning from the inputs and outputs of network. Thereafter we define the types and features of networks and vertices in a way similar to that of neural networks and neurons, and use various neural network algorithms to analyze these networks. It should be noted that if existing neural network algorithms are used, the connection structure of the network to be studied, that is different from neural networks, must be strictly defined in advance and connection weights for connected vertices only can be calculated. However, user-designed neural network models (Zhang, 2010) can be used to calculate connection weights of (user-defined) networks.

3.2.1. *Revised BP algorithm*

Backpropagation (BP) is a gradient descent algorithm and an extension of LMS algorithm. BP algorithm is a one-way propagated multilayer feedforward algorithm (Zhang and Barrion, 2006; Zhang, 2007b,c,d; Zhang and Zhang, 2008), which can be revised to determine the connection weights of multi-layer networks.

BP algorithm (Yan and Zhang, 2000; Zhang, 2010) can be revised and summarized as follows:

(1) **Perform initialization**. Choose the suitable network architecture and set all weights to smaller uniformly distributed values.
(2) **Conduct the following computation for a randomly selected sample**

    (a) **Feedforward computation**. For the vertex j of layer l, we have:

$$u_j^l(k) = \sum_{i=0}^{p} w_{ji}^l(k) y_i^{l-1}(k),$$

    where $y_i^{l-1}(k)$ is the working signal transfered from the unit i in layer $l-1$, p is the number of connections to unit j. In this study, if there is no connection between vertices i and j, then $w_{ji}^l(k) = 0$.

If the transfer function of vertex j is a sigmoid function, then

$$y_j^{l-1}(k) = 1/(1 + \exp(-u_j^{l-1}(k) + b_j^{l-1})),$$

$$\varphi'(u_j(k)) = \partial y_i^l(k)/\partial u_j(k) = y_i^l(k)(1 - y_i^l(k)).$$

If vertex j is in layer 1, i.e., $l = 1$, set $y_j^0(k) = x_j(k)$, and if vertex j is in output layer s, i.e., $l = s$, set $y_j^s(k) = O_j(k)$, and $e_j^s(k) = \hat{y}_j(k) - O_j(k)$.

(b) **Backforward computation.** For hidden vertices and output units, if the transfer function is a sigmoid function, we have:

$$\delta_j^l(k) = y_j^l(k)(1 - y_j^l(k)) \sum \delta_i^{l+1}(k) w_{ij}^{l+1}(k),$$

and

$$\delta_j^l(k) = e_j^s(k),$$

respectively. If there is no connection between vertices i and j, $w_{ij}^{l+1}(k) = 0$.

(3) **Modify weights and bias:**

$$w_{ji}^l(k + 1) = w_{ji}^l(k) + \eta \delta_j^l(k) y_i^{l-1}(k)$$

$$b_j^l(k + 1) = b_j^l(k) + \eta \delta_j^l(k)$$

If there is no connection between vertices i and j, $w_{ji}^l(k + 1) = w_{ji}^l(k) = 0$.

(4) Set $k = k + 1$, and input a new randomly selected sample, until the error

$$\sum_{i=1}^{N} \sum_{j=1}^{m} e_j^2(k)/(2mN)$$

is lower than the desired value, where m is the number of units in output layer, N is the total number of samples, and $e_j(k) = \hat{y}_j(k) - y_j(k)$.

The following are Java codes for the revised BP algorithm:

```
/*Revised BP algorithm for determining network weights and
bias. In data file, rows are samples. In a row (a sample), the
first n data are for n input vertices and the last m data are for
m output vertices. In each row of connection matrix table, the
first four values are ID of present layer, ID of vertex in present
layer, ID of next layer, ID of vertex in next layer, 1 means
connection between the two vertices. ID of input layer is 1.
Output layer has the largest ID number.*/
    public class BP {
    static int maxd,nlayer;
    static int connect[][][],n[];
    static String tablename1;
    public static void main(String[] args){
    int i,j,samp,epochs,sel;
    double h,msee;
    if (args.length!=6) System.out.println("You must input the
name of connection matrix table in the table, the name of data
table in the database, the type of transfer functions of input
and hidden vertices (1: linear function (y=x); 2: square function
(y=x^2); 3: rooted function (y=sqrt(x)); 4: sigmoid function
(y=1/(1+Math.exp(-x)))), the learning rate, desired mean square
error, and the maximum number of epochs. For example, you may
type the following in the command window: java BP bp1 bp2 4 0.05
0.0001 1000, where bp1 and bp2 are the name of connection matrix
and data table respectively, 3 means sigmoid transfer function,
0.05 is the learning rate, 1000 is the maximum number of epochs,
0.0001 is the desired mean squared error.");
    tablename1=args[0];
    String tablename2=args[1];
    sel=Integer.valueOf(args[2]).intValue();
    h=Double.valueOf(args[3]).doubleValue();
    msee=Double.valueOf(args[4]).doubleValue();
    epochs=Integer.valueOf(args[5]).intValue();
    connstr(tablename1);
    readDatabase readdata2=new
readDatabase("dataBase",tablename2,n[1]+n[nlayer]);
    samp=readdata2.m;
    double x[][]=new double[samp+1][n[1]+1];
    double y[][]=new double[samp+1][n[nlayer]+1];
    for(i=1;i<=samp;i++) {
    for(j=1;j<=n[1];j++)
```

```
    x[i][j]=(Double.valueOf(readdata2.data[i][j])).
doubleValue();
    for(j=n[1]+1;j<=n[1]+n[nlayer];j++)
    y[i][j-n[1]]=(Double.valueOf(readdata2.data[i][j])).
doubleValue(); }
    bp(sel,nlayer,n,maxd,samp,h,msee,epochs,x,y,connect); }
    public static void connstr(String tablename1) {
    int i,j,k,sum,totconn;
    readDatabase readdata1=new readDatabase("dataBase",
tablename1,5);
    totconn=readdata1.m;
    int conn[][]=new int[totconn+1][6];
    nlayer=0;
    for(i=1;i<=totconn;i++)
    for(j=1;j<=5;j++) {
    conn[i][j]=(Integer.valueOf(readdata1.data[i][j])).
intValue();
    if ((j==3) & (conn[i][j]>nlayer)) nlayer=conn[i][j]; }
    n=new int[nlayer+1];
    sum=1;
    for(i=1;i<=nlayer-1;i++) {
    n[i]=0;
    while(conn[sum][1]==i) {
    if (conn[sum][2]>n[i]) n[i]=conn[sum][2];
    sum++;
    if (sum>totconn) break; } }
    n[nlayer]=0;
    for(i=1;i<=totconn;i++)
    if ((conn[i][3]==nlayer) & (conn[i][4]>n[nlayer]))
n[nlayer]=conn[i][4];
    maxd=0;
    for(i=1;i<=nlayer;i++)
    if (n[i]>maxd) maxd=n[i];
    connect=new int[nlayer+1][maxd+1][maxd+1];
    sum=1;
    for(i=1;i<=nlayer-1;i++)
    for(j=1;j<=n[i];j++)
    for(k=1;k<=n[i+1];k++)
    if (conn[sum][4]==k) {
    connect[i][j][k]=1;
    sum++;
    if (sum>totconn) break; }
    n[0]=n[1];
```

```
   for(i=1;i<=n[0];i++)
   for(j=1;j<=n[1];j++)
   connect[0][i][j]=0;
   System.out.println("In total there are "+nlayer+"
layers in the network.");
   System.out.print("Number of vertices of each layer: ");
   for(i=1;i<=nlayer;i++)
   System.out.print(n[i]+" ");
   System.out.println();
   System.out.println("in which there are "+n[1]+" and
"+n[nlayer]+" vertices in input layer and output layer
respectively");
   System.out.println();
   System.out.println("Between-layer connection matrices
(wij=1 means there is a connection between vertices i and j; wij=0
means no connection):");
   for(i=1;i<=nlayer-1;i++) {
   System.out.println("Layer "+i+" to layer "+(i+1));
   for(j=1;j<=n[i];j++) {
   for(k=1;k<=n[i+1];k++)
   System.out.print(connect[i][j][k]+" ");
   System.out.println(); }
   System.out.println("\n"); } }
   public static double[] transfun(int sel, double x) {
   double tran[]=new double[3];
   double y=0,deri=0;
   switch(sel) {
      case 1: y=x; deri=1; break;
      case 2: y=Math.pow(x,2); deri=2*Math.sqrt(y); break;
      case 3: y=Math.sqrt(x); deri=0.5/y; break;
      case 4: y=1/(1+Math.exp(-x)); deri=y*(1-y); break; }
   tran[1]=y;
   tran[2]=deri;
   return tran; }
   public static void bp(int sel, int nlayer, int n[],
int maxd, int samp, double h, double msee, int epochs,
double x[][], double y[][], int connect[][][]) {
   int i,j,l,k,c,cs,sim,sam;
   double mse=0;
   double w[][][]=new double[nlayer+1][maxd+1][maxd+1];
   double yy[][]=new double[nlayer+1][maxd+1];
   double delta[][]=new double[nlayer+1][maxd+1];
   double u[]=new double[maxd+1];
```

```
double yyout[][]=new double[samp+1][maxd+1];
double b[][]=new double[nlayer+1][maxd+1];
int ww[]=new int[samp+1];
int cols[]=new int[samp+1];
for(l=1;l<=nlayer-1;l++)
for(j=1;j<=n[l];j++)
for(k=1;k<=n[l+1];k++)
if (connect[l][j][k]!=0) w[l+1][k][j]=Math.random();
for(l=1;l<=nlayer;l++)
for(j=1;j<=n[l];j++)
b[l][j]=Math.random();
for(sim=1;sim<=epochs;sim++) {
c=0;
for(i=1;i<=samp;i++)
ww[i]=i;
do {
cs=(int)((samp-c)*Math.random()+1);
cols[c+1]=ww[cs];
if (cs<(samp-c))
for(j=cs+1;j<=samp - c;j++)
ww[j-1]=ww[j];
c++; }
while (c<=(samp - 1));
mse=0;
for(sam=1;sam<=samp;sam++) {
//Feedforward computation
for(i=1;i<=n[0];i++)
yy[0][i]=x[cols[sam]][i];
for(l=1;l<=nlayer;l++)
for(j=1;j<=n[l];j++) {
if (l==1) {
yy[l][j]=transfun(sel,yy[l- 1][j]+b[l][j])[1];
continue; }
u[j]=0;
for(i=1;i<=n[l - 1];i++)
u[j]+=w[l][j][i]*yy[l-1][i];
if ((l>1) & (l<nlayer)) yy[l][j]=transfun(sel,u[j]+
b[l][j])[1];
else if (l==nlayer) yy[l][j]=u[j]+b[l][j]; }
//Backforward computation
for(i=1;i<=n[nlayer];i++)
delta[nlayer][i]=y[cols[sam]][i]-yy[nlayer][i];
for(l=nlayer-1;l>=2;l--) {
for(j=1;j<=n[l];j++) {
u[j]=0;
for(i=1;i<=n[l+1];i++)
```

```
    u[j]+=w[l+1][i][j]*delta[l+1][i];
    delta[l][j]=transfun(sel,yy[l][j])[2]*u[j]; } }
    //Modify weights
    for(l=2;l<=nlayer;l++)
    for(i=1;i<=n[l];i++)
    for(j=1;j<=n[l-1];j++)
    if (connect[l-1][j][i]!=0) w[l][i][j]+=h*delta[l][i]*
yy[l-1][j];
    //Modify bias
    for(l=1;l<=nlayer;l++)
    for(i=1;i<=n[l];i++)
    b[l][i]+=h*delta[l][i];
    for(i=1;i<=n[nlayer];i++)
    mse+=Math.pow(y[cols[sam]][i]-yy[nlayer][i],2)/
(2*samp*n[nlayer]);
    for(i=1;i<=n[nlayer];i++)
    yyout[cols[sam]][i]=yy[nlayer][i]; }
    if (mse<=msee) break; }
    System.out.println("Practical outputs:");
    for(i=1;i<=samp;i++) {
    for(j=1;j<=n[nlayer];j++)
    System.out.print(y[i][j]+" ");
    System.out.println(); }
    System.out.println("Simulated outputs:");
    for(i=1;i<=samp;i++) {
    for(j=1;j<=n[nlayer];j++)
    System.out.print(yyout[i][j]+" ");
    System.out.println(); }
    System.out.println();
    System.out.println("Mean squared error="+mse+"\n");
    System.out.println("Between-layer weight matrices:");
    for(l=2;l<=nlayer;l++) {
    System.out.println("Layer "+(l-1)+" to layer "+l);
    for(k=1;k<=n[l-1];k++) {
    for(j=1;j<=n[l];j++)
    System.out.print(w[l][j][k]+" ");
    System.out.println(); }
    System.out.println(); }
    System.out.println("Bias of each layer:");
    for(l=1;l<=nlayer;l++) {
    System.out.print("Layer "+l+": ");
    for(j=1;j<=n[l];j++)
    System.out.print(b[l][j]+" ");
    System.out.println(); }
    } }
```

For example, the connection structure of a network is listed below in Fig. 7:

Layer i	Vertex in layer i	Layer $i + 1$	Vertex in layer $i + 1$	Connected?
1	1	2	1	1
1	1	2	2	1
1	1	2	4	1
1	1	2	5	1
1	2	2	1	1
1	2	2	2	1
1	2	2	3	1
1	2	2	5	1
2	1	3	1	1
2	1	3	2	1
2	1	3	3	1
2	2	3	1	1
2	2	3	3	1
2	3	3	1	1
2	3	3	2	1
2	4	3	2	1
2	4	3	3	1
2	5	3	2	1
3	1	4	1	1
3	1	4	3	1
3	1	4	4	1
3	2	4	1	1
3	2	4	2	1
3	2	4	3	1
3	2	4	4	1
3	3	4	1	1
3	3	4	2	1
3	3	4	3	1
3	3	4	4	1
4	1	5	1	1
4	1	5	2	1
4	1	5	3	1
4	2	5	2	1
4	2	5	3	1
4	3	5	1	1
4	3	5	2	1
4	3	5	3	1
4	4	5	1	1
4	4	5	2	1

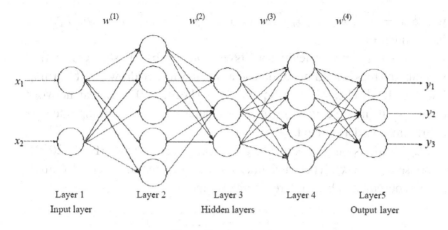

$w^{(1)}$ $w^{(2)}$ $w^{(3)}$ $w^{(4)}$

x_1

x_2

y_1
y_2
y_3

Layer 1   Layer 2   Layer 3   Layer 4   Layer5
Input layer   Hidden layers   Output layer

Figure 7.   A network.

The inputs (x_1, x_2) and outputs (y_1, y_2, y_3) of the network is listed below:

x_1	x_2	y_1	y_2	y_3
7	26	6	60	78.5
1	29	15	52	74.3
11	56	8	20	104.6
11	31	8	47	87.6
7	52	6	33	95.9
11	55	9	22	109.2
3	71	17	6	102.7
1	31	22	44	72.5
2	54	18	22	93.1
21	47	4	26	114.9
1	40	23	34	83.8
11	66	9	12	113.3
10	68	8	12	109.4

Using the revised BP algorithm, we can obtain between-layer weight matrices, mean squared error and simulated outputs.

3.2.2.   *Revised linear network algorithm*

The mathematical expression of linear network is (Zhang, 2010):

$$y = wx + b$$

where $x = (x_1, x_2, \ldots, x_n)^T$, $y = (y_1, y_2, \ldots, y_s)^T$, $b = (b_1, b_2, \ldots,$ $b_s)^T$, and $w = (w_{ij})_{s \times n}$.

Linear network may learn from its environment by adjusting connection weights and bias according to some learning law like Widrow–Hoff learning law, i.e., LMS (Least Mean Square) rule (Fecit, 2003). Linear network converges if input vectors are linearly independent and learning rate η is approximately determined.

LMS rule is only effective to one-layer linear network (Fecit, 2003). Suppose $x(k)$, $y(k)$, $\hat{y}(k)$, and $b(k)$ are network input, practical output, desired output, and bias at k respectively, and:

$$x(k) = (x_1(k), x_2(k), \ldots, x_n(k))^T$$
$$y(k) = (y_1(k), y_2(k), \ldots, y_s(k))^T$$
$$\hat{y}(k) = (\hat{y}_1(k), \hat{y}_2(k), \ldots, \hat{y}_s(k))^T$$
$$b(k) = (b_1(k), b_2(k), \ldots, b_s(k))^T$$
$$w(k) = (w_{ij}(k))$$

LMS rule is approximately a gradient descent method, expressed as the following:

$$w_{ij}(k+1) = w_{ij}(k) + \eta(y_i(k) - \hat{y}_i(k))x_j(k)$$
$$b_i(k+1) = b_i(k) + \eta(y_i(k) - \hat{y}_i(k))$$

where η is the learning rate. In this study if there is no connection between vertices i and j, $w_{ij}(k+1) = w_{ij}(k) = 0$.

The training procedure of revised linear network is explained:

(1) Calculate network output, $y = wx + b$, and error, $e = y - \hat{y}$. Here $w_{ij} = 0$, if there is no connection between vertices i and j.
(2) Compare the mean square of output error with the desired. If the error is less than the desired or the maximum number of epochs is reached, the training procedure terminates; or else continues to train.
(3) Calculate weights and bias using LMS rule and return to (1). Here $w_{ij} = 0$, if there is no connection between vertices i and j.

The Java codes for the revised linear network algorithm are listed as the following:

```
/*Revised linear network algorithm for determining network
weights and bias. In data file, rows are samples. In a row
(a sample), the first n data are for n input vertices and the last
s data are for s output vertices. In connection matrix file,
1 means that between output vertex i and input vertex j connection
exists and 0 means no connection.*/
    public class linearNet {
    public static void main(String[] args){
    int n,s,samp,epochs;
    double h,msee;
    if (args.length!=7) System.out.println("You must input the
name of connection matrix table in the table, the name of data
table in the database, the number of vertices in input layer,
the number of vertcies in output layer, the learning rate,
desired mean square error, and the maximum number of epochs. For
example, you may type the following in the command window: java
linearNet linearnet1 linearnet2 3 2 0.1 0.0001 100, where
linearnet1 and linearnet 2 are the name of connection matrix and
data table respectively, 3 and 2 are the number of vertices in
input layer and output layer respectively, 0.1 is the learning
rate, 100 is the maximum number of epochs, 0.0001 is the desired
mean squared error.");
    String tablename1=args[0];
    String tablename2=args[1];
    n=Integer.valueOf(args[2]).intValue();
    s=Integer.valueOf(args[3]).intValue();
    h=Double.valueOf(args[4]).doubleValue();
    msee=Double.valueOf(args[5]).doubleValue();
    epochs=Integer.valueOf(args[6]).intValue();
    readDatabase readdata1=new readDatabase("dataBase",
tablename1,n);
    readDatabase readdata2=new readDatabase("dataBase",
tablename2,n+s);
    samp=readdata2.m;
    double x[][]=new double[samp+1][n+1];
    double y[][]=new double[samp+1][s+1];
    int ww[][]=new int[s+1][n+1];
    for(int i=1;i<=samp;i++) {
    for(int j=1;j<=n;j++)
    x[i][j]=(Double.valueOf(readdata2.data[i][j])).
doubleValue();
```

```
    for(int j=n+1;j<=n+s;j++)
    y[i][j-n]=(Double.valueOf(readdata2.data[i][j])).
doubleValue();  }
    System.out.println("There are "+n+" input vertices and
"+s+" output vertices."+"In total "+samp+" samples are used
to traing network.");
    System.out.println("Connection matrix (wij=1 means there
is a connection between vertices i and j; wij=0 means no
connection):");
    for(int i=1;i<=s;i++)
    for(int j=1;j<=n;j++)
    ww[i][j]=(Integer.valueOf(readdata1.data[i][j])).
intValue();
    for(int i=1;i<=n;i++) {
    for(int j=1;j<=s;j++) {
    System.out.print(ww[j][i]+" ");  }
    System.out.println();  }
    System.out.println();
    linearnet(n,s,samp,h,msee,epochs,x,y,ww);  }
    public static void linearnet(int n, int s, int samp,
double h, double msee, int epochs, double x[][],
double y[][], int ww[][]) {
    int i,j,c,cs,sim,sam;
    double mse=0;
    double w[][]=new double[s+1][n+1];
    double yout[][]=new double[samp+1][s+1];
    double b[]=new double[s+1];
    double yy[]=new double[s+1];
    int p[]=new int[samp+1];
    int cols[]=new int[samp+1];
    for(i=1;i<=s;i++)
    for(j=1;j<=n;j++)
    if (ww[i][j]==1) w[i][j]=Math.random();
    else w[i][j]=0;
    for(i=1;i<=s;i++)
    b[i]=Math.random();
    for(sim=1;sim<=epochs;sim++) {
    c=0;
    for(i=1;i<=samp;i++)
    p[i]=i;
    do {
    cs=(int)((samp - c)*Math.random()+1);
    cols[c+1]=p[cs];
    if (cs<(samp - c))
    for(j=cs+1;j<=samp - c;j++)
```

```
      p[j-1]=p[j];
      c++; }
      while (c<=(samp - 1));
      mse=0;
      for(sam=1;sam<=samp;sam++) {
      for(i=1;i<=s;i++) {
      yy[i]=0;
      for(j=1;j<=n;j++)
      if (ww[i][j]==1) yy[i]+=w[i][j]*x[cols[sam]][j]+b[i];
      else yy[i]+=b[i]; }
      for(i=1;i<=s;i++)
      for(j=1;j<=n;j++)
      if (ww[i][j]==1) w[i][j]+=h*(y[cols[sam]][i]-yy[i])*
x[cols[sam]][j];
      else w[i][j]=0;
      for(i=1;i<=s;i++)
      b[i]+=h*(y[cols[sam]][i] - yy[i]);
      for(i=1;i<=s;i++)
      mse+=Math.pow(y[cols[sam]][i]-yy[i],2)/(s*samp);
      for(i=1;i<=s;i++)
      yout[cols[sam]][i]=yy[i]; }
      if (mse<=msee) break; }
      System.out.println("Practical outputs:");
      for(i=1;i<=samp;i++) {
      for(j=1;j<=s;j++)
      System.out.print(y[i][j]+" ");
      System.out.println(); }
      System.out.println("Simulated outputs:");
      for(i=1;i<=samp;i++) {
      for(j=1;j<=s;j++)
      System.out.print(yout[i][j]+" ");
      System.out.println(); }
      System.out.println();
      System.out.println("Mean squared error (MSE)="+mse+"\n");
      System.out.println("Bias vector:");
      for(i=1;i<=s;i++)
      System.out.print(b[i]+" ");
      System.out.println("\n");
      System.out.println("Weight matrix:");
      for(i=1;i<=n;i++) {
      for(j=1;j<=s;j++)
      System.out.print(w[j][i]+" ");
      System.out.println(); }
      System.out.println(); }
      }
```

3.2.3. *Probabilistic network*

Probabilistic network is a paralel realization of Bayesian classifier. A probabilistic network has two hidden layers. An exponential function is used to replace sigmoid function in the RBF network (Sprecht, 1988, 1990). In the probabilistic network, the number of pattern vertices equals to the number of trained samples and the number of summation vertices equals to the number of categories (Zhang, 2007b,c, 2010; Fig. 8).

The input of the pattern vertex in the probabilistic network is

$$g(y_i) = \exp((y_i - 1)/\sigma^2).$$

If $\sigma = c$, the network will be a Bayes classifier, where c is a constant,; if $\sigma = \infty$, it tends to be a linear classifier, and the network tends to be a neighborhood classifier if $\sigma = 0$.

The learning process of probabilistic network is:

(1) Calculate the point product of each pattern vertex: $y_i = x^T w^i$, where x is the input vector (a sample), w^i is the weight vector, and achieve the input of each pattern vertex, $g(y_i) = \exp((y_i - 1)/\sigma^2)$.
(2) Choose adequate weights and activation functions of pattern vertices such that $f_A(x)$ and $f_B(x)$ represent distribution density functions of classes A and B respectively.
(3) Calculate the weighted output.

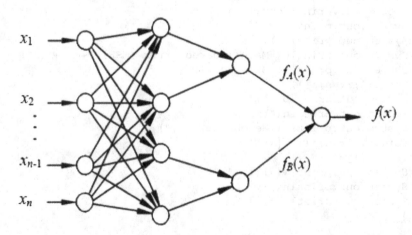

Figure 8.   Probabilistic network (Zhang, 2010).

4. Hierarchical Cluster Analysis of Networks

Classification of networks is a research subject in network analysis. It can be conducted by using cluster analysis. Hierachical cluster analysis is one of the most used cluster methods in biological research. There are many different algorithms for hierarchical cluster analysis (Zhang and Fang, 1982; Krebs, 1999; Zhang, 2007b). In the present article, a Java algorithm, including various hierarchical cluster algorithms, based on non-parametric test p of between-network difference, was developed to make cluster analysis on networks.

4.1. Algorithm

Suppose network data for ith attribute and jth network is x_{ij}, $i = 1, 2, \ldots, m$; $j = 1, 2, \ldots, n$. First, standardize the raw data with standard deviation:

$$a_{ij} = (x_{ij} - x_{bi})/s_i,$$

where,

$$x_{bi} = \sum_{j=1,\ldots,n} x_{ij}/n,$$

$$s_i = \left(\sum_{j=1,\ldots,n} (x_{ij} - x_{bi})^2/(n-1) \right)^{1/2}, \quad i = 1, 2, \ldots, m.$$

The second standardization measure is the range:

$$a_{ij} = (x_{ij} - \min x_{ik})/(\max x_{ik} - \min x_{ik}),$$
$$i = 1, 2, \ldots, m; \quad j = 1, 2, \ldots, n.$$

where $\max x_{ik} = \max(x_{i1}, x_{i2}, \ldots, x_{in})$, $\min x_{ik} = \min(x_{i1}, x_{i2}, \ldots, x_{in})$. The third measure is a ratio:

$$a_{ij} = x_{ij}/\max x_{ik},$$
$$i = 1, 2, \ldots, m; \quad j = 1, 2, \ldots, n.$$

In addition, weights can be set to each attribute, in order to denote its relative importance. The weight of ith attribute is w_i, and it should meet $w_i > 0$, $\sum_{j=1,\ldots,m} w_i = 1$. Thus the raw data will be:

$$a_{ij} = w_i a_{ij} \quad i = 1, 2, \ldots, m; \ j = 1, 2, \ldots, n.$$

Calculate between-network Euclidean distance, or Manhattan distance, or Pearson correlation, or Jaccard coefficient:

$$r_{ij} = \left(\sum_{k=1,\ldots,m} (a_{ki} - a_{kj})^2 / m \right)^{1/2},$$

$$r_{ij} = \sum_{k=1,\ldots,m} |a_{ki} - a_{kj}|/m,$$

$$r_{ij} = 1 - \sum_{k=1,\ldots,m} (a_{ki} - a_{ibar})(a_{kj} - a_{jbar}) \bigg/$$

$$\left(\sum_{k=1,\ldots,m} (a_{ki} - a_{ibar})^2 \sum_{k=1,\ldots,m} (a_{kj} - a_{jbar})^2 \right)^{1/2},$$

$$r_{ij} = (b_i + b_j)/(c_i + c_j - e), \quad i, j = 1, 2, \ldots, n,$$

where b_i is the non-zero number present at network i but not at network j, b_j is the non-zero number present at network j but not at network i, c_i and c_j are the non-zero number at network i and network j respectively, and e is the non-zero number shared by network i and network j.

Using the non-parametric randomization statistic test (Zhang and Barrion, 2006; Zhang, 2011a), statistic p value for between-network difference may be calculated. A smaller p value means a greater difference between two networks. Thus take $r = 1 - p$ as between-network distance, and the distance between networks i and j is $r_{ij} = 1 - p_{ij}$.

Let $R = (r_{ij}) = (1 - p_{ij})$. Each network makes a cluster. There are nine algorithms to define between-cluster distance, i.e., the ones for shortest distance, longest distance, median distance, weight center, cluster averaging, changeable distance, changeable cluster averaging, corrected cluster averaging, and deviation square sum (Zhang and Fang, 1982; Krebs, 1999; Zhang, 2007b). Between-cluster distance for these algorithm is defined as shortest distance between networks, longest distance between networks, median distance between networks with beta parameter, distance between weight centers and each of which is the mean of networks in same cluster, square root of average square distance of any two networks between two clusters, skewed median distance with a beta parameter, changeable cluster averaging with a beta parameter, corrected cluster averaging with a beta parameter, and deviation square sum.

For median distance assume that two clusters are P and Q, r_{ij} is the shortest distance between P and Q, $i \in P$, $j \in Q$, r_{kl} is the longest distance between P and Q, $k \in P$, $l \in Q$, r_{st} is the longest distance within P, $s \in P$, $t \in P$, and r_{uv} is the longest distance within Q, $u \in Q$, $v \in Q$. The distance between P and Q is:

$$(r_{ij}^2/2 + r_{kl}^2/2 + \beta^*(r_{st}^2 + r_{uv}^2))^{0.5}.$$

For changeable distance, the distance between P and Q is:

$$((1-\beta)^*r_{ij}^2/2 + (1-\beta)^*r_{kl}^2/2 + \beta^*(r_{st}^2 + r_{uv}^2))^{0.5}.$$

For changeable cluster averaging, the distance between P and Q is:

$$(1-\beta)/(n_p n_q) \sum_{i \in P, j \in Q} r_{ij}^2 + \beta^* \left(\sum_{i \in P, j \in P} r_{ij}^2/n_p + \sum_{i \in Q, j \in Q} r_{ij}^2/n_q \right).$$

For corrected cluster averaging, the distance between P and Q is:

$$1/(n_p n_q) \sum_{i \in P, j \in Q} r_{ij}^2 + \beta^* \left(\sum_{i \in P, j \in P} r_{ij}^2/n_p + \sum_{i \in Q, j \in Q} r_{ij}^2/n_q \right).$$

where n_p and n_q are the number of networks in P and Q respectively.

For deviation square sum, assume the square sum of deviation for cluster P and Q are p and q respectively. Combine P and Q into a new cluster R, the square sum of deviation for cluster R is r, then the distance between P and Q is:

$$r - p - q.$$

Select the clusters with smallest distance, and combine them into a new cluster. By combining each clustering, the number of clusters will be decreased, until all networks are combined into the same cluster.

If between-cluster distance, $r = 1 - p$, is greater than the significance degree, i.e., 0.99, or 0.95, then the two clusters are significantly different.

The algorithm, HierarClusterAnal, is implemented as a Java program based on JDK 1.1.8, in which several clusters and an HTML file is included. In cluster data file, the first row is attribute ID Number, and the first column is network ID Number. If weight of attribute is considered then the first row is weight of each attribute.

Some of the Java codes for the algorithm are:

```
for(i=1;i<=m;i++) {
ww[i]=(Double.valueOf(sp.substring(0,sp.indexOf(' ')))).
    doubleValue();
sp=(sp.substring(sp.indexOf(' '))).trim(); }
for(i=1;i<=n;i++) {
dum=(Double.valueOf(sp.substring(0,sp.indexOf(' ')))).
    doubleValue();
sp=(sp.substring(sp.indexOf(' '))).trim();
for(j=1;j<=m;j++) {
if (!((i==n) & (j==m)))
a[i][j]=(Double.valueOf(sp.substring(0,sp.indexOf(' ')))).
    doubleValue();
else {a[i][j]=(Double.valueOf(sp)).doubleValue(); break; }
sp=(sp.substring(sp.indexOf(' '))).trim(); } }
rantest=new RandTest();
if (sta==1)
for(i=1;i<=m;i++) {
h[i]=0;
p[i]=0;
for(j=1;j<=n;j++) p[i]+=a[j][i];
p[i]/=n;
for(k=1;k<=n;k++) h[i]+=Math.pow(a[k][i]-p[i],2);
h[i]=Math.sqrt(h[i]/(n-1));
for(j=1;j<=n;j++) a[j][i]=(a[j][i]-p[i])/h[i]; }
else if (sta==2)
for(i=1;i<=m;i++) {
aa=-1e+100;
bb=1e+100;
for(j=1;j<=n;j++) {
if (a[j][i]>aa) aa=a[j][i];
if (a[j][i]<bb) bb=a[j][i]; }
for(j=1;j<=n;j++) a[j][i]=(a[j][i]-bb)/(aa-bb); }
else if (sta==3)
for(i=1;i<=m;i++) {
aa=-1e+100;
for(j=1;j<=n;j++)
if (a[j][i]>aa) aa=a[j][i];
for(j=1;j<=n;j++) a[j][i]=a[j][i]/aa; }
if (selw==1) {
dum=0;
for(i=1;i<=m;i++) dum+=ww[i];
if (Math.abs(dum-1)>1e-05)
for(i=1;i<=m;i++) ww[i]/=dum;
for(i=1;i<=m;i++)
```

```
for(j=1;j<=n;j++) a[j][i]*=ww[i]; }
if (choice==9) {
bb1=1;
u[bb1]=0;
nu[bb1]=n;
for(i=1;i<=nu[bb1];i++) x[bb1][i]=i;
for(i=1;i<=nu[bb1];i++) y[bb1][i]=1;
class_distance(); }
else {
for(i=1;i<=n-1;i++)
for(j=i+1;j<=n;j++) {
sampledistance();
r[j][i]=r[i][j]; }
for(i=1;i<=n;i++) r[i][i]=0;
for(i=1;i<=n;i++)
for(j=1;j<=n;j++) r1[i][j]=r[i][j]; }
classification();
order();
String stt="";
xmax=n;
umax=u[bb1];
k=1;
for(i=1;i<=nu[1];i++)
for(j=1;j<=n;j++) if (y[1][j]==i) {vs[k]=String.valueOf(j);
k++;}
new GraphicsFrame(new
ClusterGraphics(vs,bb1,n,xmax,umax,u,nu,y),stt).
resize(710,560);
String iss="";
for(k=1;k<=bb1;k++) {
iss+="r=1-p="+(int)(u[k]*10000)/10000.00+"\n";
for(i=1;i<=nu[k];i++) {
iss+="(";
for(j=1;j<=n;j++) if (y[k][j]==i) iss+=j+" ";
iss+=") "; }
iss+="\n"; }
output.editt1.appendText(iss);
output.editt1.appendText("---------------------------------
------------------------------------\n");
public double sampledistance() {
for(k=1;k<=m;k++) {
aaa1[k]=(int)(a[i][k]*100+0.5)/100.0;
aaa2[k]=(int)(a[j][k]*100+0.5)/100.0; }
r[i][j]=1-rantest.randTest(m,aaa1,aaa2,sim,sele);
return r[i][j]; }
```

```
public double[][] class_distance() {
if ((choice==1) | (choice==2)) {
for(i=1;i<=nu[bb1]-1;i++)
for(j=i+1;j<=nu[bb1];j++) {
if (choice==1) r[i][j]=1e+100;
if (choice==2) r[i][j]=  - 1e+100;
for(k=1;k<=n;k++)
if (x[bb1][k]==i) {
for(int kk=1;kk<=n;kk++)
if (x[bb1][kk]==j)
{if ((choice==1) & (r1[k][kk]<r[i][j])) r[i][j]=r1[k][kk];
if ((choice==2) & (r1[k][kk]>r[i][j])) r[i][j]=r1[k][kk];}}}}
if ((choice==3) | (choice==6)) {
int a1=0,a2=0,a3=0,a4=0;
double cc=1e+100;
for(i=1;i<=nu[bb1]-1;i++)
for(j=i+1;j<=nu[bb1];j++) {
r[i][j]=0;
for(int ii=1;ii<=2;ii++) {
if (ii==1) cc=1e+100; else if (ii==2) cc=  - 1e+100;
for(k=1;k<=n;k++)
if (x[bb1][k]==i) {
for(int kk=1;kk<=n;kk++)
if (x[bb1][kk]==j)
{if ((ii==1) & (r1[k][kk]<cc)) cc=r1[k][kk];
if ((ii==2) & (r1[k][kk]>cc)) cc=r1[k][kk]; } }
if (choice==3) r[i][j]+=1/2.0*Math.pow(cc,2);
else if (choice==6) r[i][j]+=(1-be)/2.0*Math.pow(cc,2); }
for(int ii=1;ii<=2;ii++) {
int jj;
if (ii==1) jj=i; else jj=j;
cc=-1e+100;
for(k=1;k<=n;k++)
if (x[bb1][k]==jj) {
for(int kk=1;kk<=n;kk++)
if (x[bb1][kk]==jj)
{if (r1[k][kk]>cc)
{cc=r1[k][kk];
if (ii==1)
{a1=k;
a2=kk; }
else
{a3=k;
a4=kk; } } } } }
r[i][j]=Math.sqrt(r[i][j]+be*(Math.pow(r1[a1][a2],2)+
   Math.pow(r1[a3][a4],2))); } }
```

```
if (choice==4) {
for(i=1;i<=nu[bb1];i++)
for(j=1;j<=m;j++) {
a1[i][j]=0;
for(k=1;k<=n;k++) if (x[bb1][k]==i) a1[i][j]+=a[k][j];
a1[i][j]/=y[bb1][i]; }
for(i=1;i<=n;i++)
for(j=1;j<=m;j++) a2[i][j]=a[i][j];
for(i=1;i<=nu[bb1];i++)
for(j=1;j<=m;j++) a[i][j]=a1[i][j];
for(i=1;i<=nu[bb1]-1;i++)
for(j=i+1;j<=nu[bb1];j++) sampledistance();
for(i=1;i<=n;i++)
for(j=1;j<=m;j++) a[i][j]=a2[i][j]; }
if ((choice==5) | (choice==7) | (choice==8)) {
for(i=1;i<=nu[bb1];i++)
for(j=i+1;j<=nu[bb1];j++) {
r[i][j]=0;
for(k=1;k<=n;k++)
if (x[bb1][k]==i) {
for(int kk=1;kk<=n;kk++)
if (x[bb1][kk]==j) r[i][j]+=Math.pow(r1[k][kk],2); }
r[i][j]=Math.sqrt(r[i][j]/(y[bb1][i]*y[bb1][j]));
if ((choice==7) | (choice==8)) {
double ww=(1-be)*r[i][j];
double ff=r[i][j];
double dd=0;
for(int ii=1;ii<=2;ii++) {
int jj;
double cc=0;
if (ii==1) jj=i; else jj=j;
for(k=1;k<=n;k++)
if (x[bb1][k]==jj) {
for(int kk=1;kk<=n;kk++)
if (x[bb1][kk]==jj) cc+=Math.pow(r1[k][kk],2); }
dd+=cc/y[bb1][jj]; }
if (choice==7) r[i][j]=ww+be*dd;
else if (choice==8) r[i][j]=ff+be*dd; } } }
if (choice==9) {
for(i=1;i<=nu[bb1];i++)
for(j=1;j<=m;j++) {
a1[i][j]=0;
for(k=1;k<=n;k++)
if (x[bb1][k]==i) a1[i][j]+=a[k][j];
a1[i][j]/=y[bb1][i]; }
for(i=1;i<=nu[bb1];i++) {
```

```
q[i]=0;
for(j=1;j<=m;j++) {
for(k=1;k<=n;k++)
if (x[bb1][k]==i) q[i]+=Math.pow(a[k][j]-a1[i][j],2); } }
for(i=1;i<=nu[bb1]-1;i++)
for(j=i+1;j<=nu[bb1];j++) {
r[i][j]=0;
for(int kk=1;kk<=m;kk++) {
h[kk]=0;
for(k=1;k<=n;k++)
if ((x[bb1][k]==i) | (x[bb1][k]==j)) h[kk]+=a[k][kk];
h[kk]/=y[bb1][i]+y[bb1][j]; }
double cc=0;
for(int kk=1;kk<=m;kk++) {
for(k=1;k<=n;k++)
if ((x[bb1][k]==i) | (x[bb1][k]==j))
   cc+=Math.pow(a[k][kk]-h[kk],2); }
r[i][j]=cc-q[i]-q[j]; } }
return r; }

public void classification() {
bb1=1;
u[bb1]=0;
nu[bb1]=n;
for(i=1;i<=nu[bb1];i++) x[bb1][i]=i;
for(i=1;i<=nu[bb1];i++) y[bb1][i]=1;
do {
aa=1e+200;
for(i=1;i<=nu[bb1]-1;i++)
for(j=i+1;j<=nu[bb1];j++)
if (r[i][j]<=aa) aa=r[i][j];
aa1=0;
for(i=1;i<=nu[bb1]-1;i++)
for(j=i+1;j<=nu[bb1];j++)
if (Math.abs(r[i][j]-aa)<=1e-05)
{aa1++;
v[aa1]=i;
w[aa1]=j;}
for(i=1;i<=nu[bb1];i++) s[i]=0;
nn1=0;
for(i=1;i<=aa1;i++)
if ((v[i]!=0) & (w[i]!=0))
{nn1++;
for(j=1;j<=aa1;j++)
if ((v[j]==v[i]) | (v[j]==w[i]) | (w[j]==w[i])
   | (w[j]==v[i]))
```

```
{s[v[j]]=nn1;
s[w[j]]=nn1;
if (j!=i) {v[j]=0;
w[j]=0;} }
v[i]=0;
w[i]=0; }
for(i=1;i<=nn1;i++) {
y[bb1+1][i]=0;
for(j=1;j<=nu[bb1];j++)
if (s[j]==i) {
for(k=1;k<=n;k++)
if (x[bb1][k]==j) x[bb1+1][k]=i;
y[bb1+1][i]+=y[bb1][j]; } }
for(i=1;i<=nu[bb1];i++)
if (s[i]==0) {
nn1++;
for(k=1;k<=n;k++)
if (x[bb1][k]==i) x[bb1+1][k]=nn1;
y[bb1+1][nn1]=y[bb1][i]; }
bb1++;
u[bb1]=aa;
nu[bb1]=nn1;
class_distance(); }
while (nu[bb1]>1); }

public int[][] order() {
for(k=1;k<=n;k++) y[bb1][k]=1;
for(i=bb1-1;i>=1;i--) {
int rr=0;
for(j=1;j<=nu[i+1];j++) {
int ww=0;
for(k=1;k<=n;k++)
if (y[i+1][k]==j) {ww++;
v[ww]=k;}
int vv=0;
for(int ii=1;ii<=ww;ii++) {
int ee=0;
for(int jj=ii-1;jj>=1;jj--) {
if (x[i][v[ii]]==x[i][v[jj]])
{y[i][v[ii]]=y[i][v[jj]];
break;}
ee++; }
if (ee==ii-1) {vv++;
y[i][v[ii]]=rr+vv; } }
rr+=vv; } }
return y; }
```

4.2. *Application*

There were 14 arthropod networks with six attributes, as indicated in Table 1.

Use the HierarClusterAnal algorithm, and choose cluster averaging as the cluster algorithm and Euclidean distance as between-network distance. Data were standardized with ratio standardization, and non-weighting for attributes. Set 100 randomizations. The results of hierarchical cluster analysis of networks are listed (also see Fig. 9):

$r = 1 - p = 0.0$
(1) (2) (3) (8) (9) (10) (11) (4) (13) (6) (7) (5) (12) (14)
$r = 1 - p = 0.02$
(1) (2) (3) (8) (9) (10) (11) (4) (13) (6 7) (5) (12) (14)
$r = 1 - p = 0.12$
(1) (2) (3) (8) (9) (10) (11) (4) (13) (6 7) (5 12) (14)
$r = 1 - p = 0.3$
(1) (2) (3) (8) (9) (10) (11) (4 13) (6 7) (5 12) (14)
$r = 1 - p = 0.3299$
(1) (2) (3) (8) (9 10) (11) (4 13) (6 7) (5 12) (14)
$r = 1 - p = 0.58$
(1) (2 3) (8) (9 10) (11) (4 13) (6 7) (5 12) (14)
$r = 1 - p = 0.6053$
(1) (2 3) (8) (9 10) (11) (4 13) (6 7) (5 12 14)
$r = 1 - p = 0.7201$
(1) (2 3) (8) (9 10) (11) (4 6 7 13) (5 12 14)
$r = 1 - p = 0.8139$
(1) (2 3) (8) (9 10 11) (4 6 7 13) (5 12 14)
$r = 1 - p = 0.9802$
(1) (2 3) (8 9 10 11) (4 6 7 13) (5 12 14)
$r = 1 - p = 0.9934$
(1) (2 3) (8 9 10 11) (4 5 6 7 12 13 14)
$r = 1 - p = 0.995$
(1 2 3) (8 9 10 11) (4 5 6 7 12 13 14)
$r = 1 - p = 0.9983$
(1 2 3 8 9 10 11) (4 5 6 7 12 13 14)
$r = 1 - p = 0.9991$
(1 2 3 4 5 6 7 8 9 10 11 12 13 14)

Table 1. Fourteen arthropod networks with six attributes.

ID	Country	Data set	Taxa	No. Taxa	Mean pure LCC	Number of statistcially significant interactions ($p <= 0.01$)			
						Total (n)	n/N	Positive interactions (w)	w/n
1	China	06 Sep	Functional group	4	0.2032	1	0.167	1	1.000
2	China	06 Sep	Functional group	4	0.1804	2	0.333	2	1.000
3	China	06 Oct	Functional group	4	0.1285	2	0.333	2	1.000
4	Philippines	Mar	Functional group	21	0.0029	3	0.014	2	0.667
5	Philippines	Apr	Functional group	20	0.0151	12	0.063	10	0.833
6	Philippines	Sep	Functional group	21	0.0259	4	0.019	4	1.000
7	Philippines	Oct	Functional group	21	0.0178	3	0.014	3	1.000
8	Philippines	Mar	Macro functional group	7	−0.0012	2	0.095	1	0.500
9	Philippines	Apr	Macro functional group	7	0.0952	2	0.095	2	1.000
10	Philippines	Sep	Macro functional group	7	0.0963	1	0.048	1	1.000
11	Philippines	Oct	Macro functional group	7	0.0508	2	0.095	2	1.000
12	China	06 Sep	Family	23	0.0276	12	0.047	10	0.833
13	China	06 Oct	Family	23	0.0082	7	0.028	7	1.000
14	China	06 Oct	Family	27	0.0171	15	0.043	12	0.800

N: total possible links.

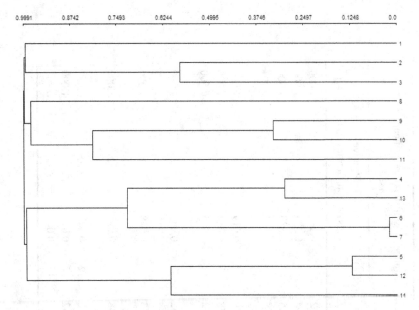

Figure 9. A hierarchical cluster analysis of networks (cluster averaging, Euclidean distance, data standardized with ratio standardization, and non-weighting).

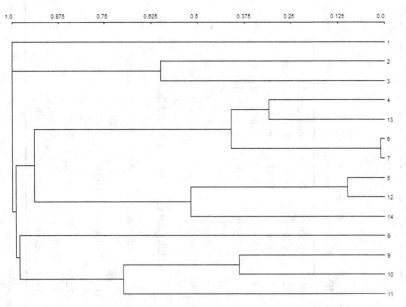

Figure 10. A hierarchical cluster analysis of networks (shortest distance, Euclidean distance, data standardized with ratio standardization, and non-weighting).

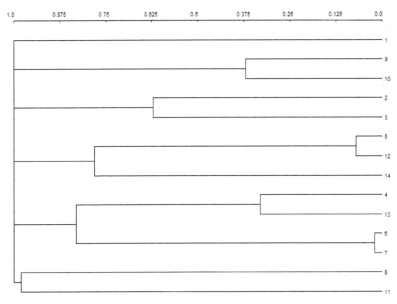

Figure 11. A hierarchical cluster analysis of networks (longest distance, Euclidean distance, data standardized with ratio standardization, and non-weighting).

If we set the significance degree as 0.95 (95% confidence degree), then from results of the hierarchical cluster analysis of networks, these cluster patterns are better choices:

(1) (2 3) (8 9 10 11) (4 6 7 13) (5 12 14)
(1) (2 3) (8 9 10 11) (4 5 6 7 12 13 14)
(1 2 3) (8 9 10 11) (4 5 6 7 12 13 14)
(1 2 3 8 9 10 11) (4 5 6 7 12 13 14)

Choose shortest and longest distance respectively and let the other choices remain unchanged. The results are indicated in Fig. 10 and Fig. 11.

Generally the results for cluster averaging are better than other cluster algorithms.

❧ CHAPTER 8 ❧

Complex Networks and Network Analysis

1. Complex Networks

Up till now a lot of research works have been done for simple networks. Many results and methods have been obtained from these researches. However, the network systems met in the recent years have become more and more complex. Occasionally there are millions of vertices and edges in a complex network. It will be impossible to approach such networks by using classical methods or algorithms.

Fortunately, graph theory, combinatorial optimization, statistics, and stochastic processes are quickly becoming the scientific basis and effective tools for studying complex networks.

In general, complex networks can be divided into four categories, namely social networks, information networks, technological networks and biological networks:

(1) social networks reflect the interaction patterns of human population. Social networks include friend networks, business networks, family networks, World Wide Web (www), etc.
(2) information networks refer to literature indexing networks, computer networks, etc.

(3) technological networks include electrical systems, power networks, energy grids, transportation networks, banking networks, telephone networks, internet, etc.
(4) biological networks include metabolic networks, protein interaction networks, gene regulatory networks, food webs, neural networks, the blood circulatory system, etc.

1.1. *Measurement of complex networks*

Watts and Strogatz (1998) found that a complex network may be measured by two topological indices, clustering coefficient (average number of adjacent edges of a vertex, C) and length of the characteristic path (the median of average shortest distances between two vertices, L). A complex system with shorter characteristic path and larger clustering coefficient is considered as a small world network (Fig. 1). The small world network is an important tool to approach complex systems.

Suppose the total number of vertices is n, the number of adjacent vertices of vertex i is n_i, then the clustering coefficient of the network is

$$C = 1/n \sum_{i=1}^{n} C_i,$$

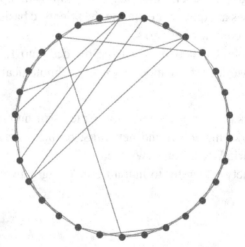

Figure 1. A small world network (clustering coefficient = 0.185, length of characteristic path = 2.485, connection probability = 0.34).

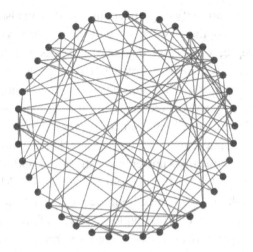

Figure 2. A complete random graph (clustering coefficient = 0.05, length of the characteristic path = 2.679, connection probability = 1).

where C_i is the clustering coefficient of vertex i,

$$C_i = 2/[n_i(n_i - 1)] \sum_{j=1}^{ni} d_j,$$

$d_i = 1$, if $e_{ij} \in E_i$, or else, $d_i = 0$; where $0 \le C \le 1$.

For a vertex, v, and an arbitrary vertex, j, the distance between the two vertices is d_{vj}. For d_{vj} of each vertex v, in the network, calculate the average d_v. The length of the characteristic path, L, is the median of all d_v's.

Clustering coefficient and the length of characteristic path should be divided by the corresponding values of the complete network (i.e., the network in which every vertices have the same number of adjacent edges) with the same size (i.e., the same number of vertices and edges) to stabilize the network.

Random graph is the simplest network (Fig. 2). In the formation of Erdos–Renyi random graph, each time a pair of n vertices are randomly (uniformly distributed) connected and added to the graph (Erdos and Renyi, 1959). The random graph demonstrates small world effect and is thus an approximation of the real world.

As a connectivity property of graph, connectivity can also be used as a performance indicator of the network. Assume the proportion of the number of vertices with degree k is p_k, the degree of vertex is therefore a random variable. The probability distribution of degree is called degree distribution. Degree distribution represents the connection degree of a complex network.

In the random network, degree distribution is a binomial distribution, and its limit model is Poisson distribution. In a random network, the majority of vertices have the same degree with the average. In the complex network, degree distribution is typically a power law distribution, and such a network is called a scale-free network (Barabasi and Albert, 1999). Many complex networks, such as the internet, metabolic networks, communication networks, etc., have their degree distributions as that of scale-free network and are thus scale-free networks. The scale-free property is an important feature of the complex networks. Researchers are trying to approach the causes and mechanisms of the property in recent years. The possible causes and mechanisms include the following: (1) with the addition of new vertices, the network continues to expand; (2) new vertices tend to connect to already better connected vertices (Barabasi and Albert, 1999).

Some complex networks are exponentially degree-distributed, such as energy grids, railway networks, while others are complex networks with the exponent-power law degree distribution, such as cooperation networks, etc.

According to the research, the average length of the network paths is almost unchanged if some vertices are randomly removed from the complex network. However, it will grow rapidly as vertices are targetedly removed.

The study of topology of complex networks will help to facilitate deep understanding of the network dynamics, for example, the influence of species loss on ecosystems. Although the dynamics of complex networks is extremely important, however, the research is still seldom used due to lack of relevant knowledge. In general, at present there are not enough mathematical tools, methods and performance indicators to treat complex networks (Zhang, 2007b). Further studies on network dynamics and behaviors are required.

1.2. Generation and calculation of complex networks

1.2.1. Generation of random graphs

The method to generate a Erdos–Renyi random graph is that for *n* vertices each time a pair of vertices are randomly connected and added to the graph. The following are Java codes, randGraphER, for generating a Erdos–Renyi random graph:

```
/*Erdos—Renyi algorithm to generate a random graph.*/
/*v: number of vertices; e: number of edges to be added; d[1—v][1
—v]: adjacency matrix of the generated random graph.*/
public class randGraphER {
public static void main(String[] args){
int v, e;
if (args.length!=2)
System.out.println("You must input the number of vertices and
the number of edges. For example, you may type the following in the
command window: java randGraphER 20 100, where 20 is the number of
vertices and 100 is the number of edges.");
v = Integer.valueOf(args[0]).intValue();
e = Integer.valueOf(args[1]).intValue();
randGraph(v,e); }
public static int[][] randGraph(int v, int e) {
int i,j,k,a1,a2;
int d[][] = new int[v + 1][v + 1];
for(i=1;i<=v;i++)
for(j=1;j<=v;j++) d[i][j] = 0;
for(k=1;k<=e;k++) {
do {
a1 = (int)Math.round(v*Math.random()+0.5);
a2 = (int)Math.round(v*Math.random()+0.5); }
while ((a1==a2)|(d[a1][a2]==1));
d[a1][a2]=1;
d[a2][a1]=1; }
System.out.println("Random graph (between-vertex edges):");
for(i = 1;i<=v—1;i++)
for(j = i + 1;j<=v;j++)
if (d[i][j]==1) System.out.println(i+"—"+j);
return d; }
}
```

The method of Watts and Strogatz (1998), for generating a random graph, is,

(1) produce a complete graph, in which each vertex has k adjacent edges;
(2) for each vertex, the edge is re-oriented with the probability p; this generates a complete random graph if $p = 1$; if $0 < p < 1$, the generated graph is likely similar to a small world network.

The following are Java codes of Watts and Strogatz algorithm for generating a random graph:

```
/*Watts—Strogatz algorithm to generate a random graph.*/
/*v: number of vertices; m: number of edges to be added; p:
probability for generating edges. d[1—v][1—v]: adjacency
matrix of the generated random graph.*/
public class randGraphWS {
public static void main(String[] args){
int v, e;
double p;
if (args.length!=3)
System.out.println("You must input the number of vertices, the
number of edges and the probability. For example, you may type
the following in the command window: java randGraphWS 20 200
0.3, where 20 is the number of vertices and 200 is the number of
edges, 30.3 is the probability.");
v = Integer.valueOf(args[0]).intValue();
e = Integer.valueOf(args[1]).intValue();
p = Double.valueOf(args[2]).doubleValue();
randGraph(v, e, p); }
public static int[][] randGraph(int v, int e, double p) {
int i,j,k,a1,a2;
int d[][] = new int[v+1][v+1];
for(i=1;i<=v;i++)
for(j=1;j<=v;j++)
if (i==j) d[i][j]=0;
else if ((j==(i+1))|(i==(j+1))) d[i][j] = 1;
else d[i][j] = 0;
d[v][1] = 1;
for(k=1;k<=e;k++) {
do {
a1 = (int)Math.round(v*Math.random()+0.5);
a2 = (int)Math.round(v*Math.random()+0.5); }
while ((a1==a2)|(d[a1][a2]==1));
if (Math.random()<=p) {
d[a1][a2] = 1;
d[a2][a1] = 1; }
```

```
else {
d[a1][a2] = 0;
d[a2][a1] = 0; } }
System.out.println("Random graph (between-vertex edges):");
for(i=1;i<=v-1;i++)
for(j=i+1;j<=v;j++)
if (d[i][j]==1) System.out.println(i+"-"+j);
return d; }
}
```

1.2.2. Generation of scale-free graphs

Barabasi and Albert (1999) presented a method to generate scale-free random graphs. Suppose there are n vertices. For every pair of vertices i and j, an edge is generated with the probability, p. When a vertex is connected and added, all edges are then generated in the way similar to the generation of random graph. New vertex traverses all existing vertices, but is connected and added with a fixed probability. However the probability values are different. The drawback of this algorithm is that the vertex connected and added earlier tends to be the attractor more likely, which is different from the real world. To this end, they gave a fitness value for each vertex. The rate of a vertex for accepting a new connection is positively proportional to its fitness value. At the start of the calculation, each vertex is assigned a uniform distributed fitness value. When the vertex is assigned a new edge, it increases the fitness value. Thus, with the increase of the connected edges the fitness value grows and therefore a positive feedback loop is formed.

The following are Java codes, randGraphScaFr, for generating scale-free random graph:

```
/*Barabasi-Albert algorithm to generate a scale-free random
graph.*/
/*v: number of vertices; m: number of edges to be added; p:
probability for generating edges; d[1-v][1-v]: adjacency
matrix of the generated scale-free random graph.*/
public class randGraphScaFr {
public static void main(String[] args){
int v, e;
double p;
if (args.length!=3)
System.out.println("You must input the number of vertice, the
number of edges and the probability. For example, you may type
```

the following in the command window: java randGraphScaFr 20
200 0.3, where 20 is the number of vertices and 200 is the
number of edges, 0.3 is the probability.");

```
v = Integer.valueOf(args[0]).intValue();
e = Integer.valueOf(args[1]).intValue();
p = Double.valueOf(args[2]).doubleValue();
randGraph(v,e,p); }
public static int[][] randGraph(int v, int e, double p) {
int i, j, k, a1, a2;
double b;
int d[][] = new int[v+1][v+1];
double s[] = new double[v+1];
s[0] = 0;
for(i=1; i<=v; i++)
s[i] = s[i-1]+Math.random();
for(k=1;k<=e;k++) {
b = s[v]*Math.random();
for(i=0;i<=v-1;i++)
if ((b>=s[i]) & (b<s[i+1])) {
s[i+1]+=Math.random();
do
a1 = (int)Math.round(v*Math.random()+0.5);
while (((i+1)==a1)|(d[i+1][a1]==1));
if (Math.random()<=p) {
d[i+1][a1] = 1;
d[a1][i+1] = 1; }
else {
d[i+1][a1] = 0;
d[a1][i+1] = 0; }
break; } }
System.out.println("Scale-free random graph (between-vertex
edges):");
for(i=1;i<=v-1;i++)
for(j=i+1;j<=v;j++)
if (d[i][j]==1) System.out.println(i+"-"+j);
return d; }
}
```

1.2.3. *Generation of complex networks*

Cancho and Sole (2001) algorithm can generate a variety of complex networks with diverse degree distributions. Optimize the energy function $E(\lambda)$:

$$E(\lambda) = \lambda d + (1 - \lambda)\rho,$$
$$\rho = m/(n - 1),$$

where $0 \leq \lambda \leq 0.25$, and m is the average degree of vertices. Reorganize the random network and change λ. As λ increases, the resulting network will be random network, exponential network, and scale-free network. If $\lambda > 0.25$, the resulting network is a star network.

1.2.4. *Degree distribution and network types*

As mentioned earlier, a network can be random, exponential, or scale-free network. The degree distribution of network can be determined by binomial distribution, Poisson distribution, exponential distribution, and power law distribution.

In a random network, the majority of vertices have the same degree as the average. The coefficient of variation, H, can be used to describe the type of a network (Zhang, 2007b; Zhang and Zhan, 2011),

$$H = s^2/u,$$
$$\bar{u} = \sum d_i/v,$$
$$s^2 = \sum (d_i - \bar{u})^2/(v - 1),$$

where $\bar{u}$, s^2 are the mean and variance of network degree; v is the number of vertices and d_i is the degree of vertex i, where $i = 1, 2, \ldots, v$. The network is a random network, if $H \leq 1$. Calculate $\chi^2 = (v - 1)H$, and if $\chi^2_{1-\alpha}(v-1) < \chi^2 < \chi^2_{\alpha}(v-1)$, the network is a complete random network. It is a complex network, if $H > 1$, and to some extent, network complexity increases with H.

We define E, $E = s^2 - \bar{u}$, as the entropy of network (Zhang and Zhan, 2011). A more complex network has the larger entropy. If $E \leq 0$ the network is a random network and it is a complex network if $E > 0$.

In addition, the type of a network can be determined by using the following indicator (Zhang, 2007b; Zhang and Zhan, 2011):

$$H = v^* \sum d_i(d_i - 1)/\left[\sum d_i \left(\sum d_i - 1\right)\right].$$

The network is a random network, if $H \approx 1$. Calculate $\chi^2 = H(\sum d_i - 1) + v - \sum d_i$, and if $\chi^2 < \chi^2_{\alpha}(v-1)$, the network is a complete random network. It is a complex network if $H > 1$, and network complexity increases with H.

The skewness S, of degree distribution can be used to measure the degree of skewness in relative to the symmetric distribution, for example, the normal distribution ($S = 0$) (Sokal and Rohlf, 1995):

$$S = v \sum (d_i - \bar{u})^2 / [(v - 1)(v - 2)s^3].$$

The following are Java codes, netType, for calculating the degree distribution and network type:

```
/*v: number of vertices; d[1—v][1—v]: adjacency matrix to
reflect the feature of edges, e.g., dij = dji=0 means no edge
between vertices i and j; dij=-dji, and | dij|=1,~means there is
an edge between vertices i and j; dij = dji = 2, means there are
parallel edges between vertices i and j; dii = 3 means there is
a self-loop for vertex i; dii = 4 means isolated vertex; dii=5
means isolated vertex i with self-loop. */
public class netType {
public static void main(String[] args){
int i, j, v, n;
if (args.length!=1)
System.out.println("You must input the name of table in the
database. For example, you may type the following in the command
window: java netType nettype, where nettype is the name of table.
Graph is stored in the table using two arrays listing and was
transformed to adjacency matrix by method adjMatTwoArr.");
String tablename=args[0];
readDatabase                                    readdata=new
readDatabase("dataBase",tablename, 3);
n = readdata.m;
int a[] = new int[n+1];
int b[] = new int[n+1];
int c[] = new int[n+1];
int d[][] = new int[n+1][n+1];
for(i = 1;i<=n;i++) {
a[i] = (Integer.valueOf(readdata.data[i][1])).intValue();
b[i] = (Integer.valueOf(readdata.data[i][2])).intValue();
c[i] = (Integer.valueOf(readdata.data[i][3])).intValue(); }
adjMatTwoArr adj = new adjMatTwoArr();
adj.dataTrans(a,b,c);
v = adj.v;
for(i=1;i<=v;i++)
for(j=1;j<=v;j++) d[i][j]=adj.d[i][j];
netType(v,d); }
public static void netType(int v, int d[][]) {
int i, j, k, l, m, rr, ty, r;
```

```
double it,pp,ss,qq,k1,k2,chi,mean,var,hr,h,skew;
int deg[] = new int[v+1];
int p[] = new int[v+1];
double fr[] = new double[v+1];
double pr[] = new double[v+1];
for(i=1;i<=v;i++) {
deg[i] = 0;
for(j=1;j<=v;j++) {
if (Math.abs(d[i][j])==1) deg[i]++;
if ((d[i][j]==2)|(d[i][j]==3)|(d[i][j]==5)) deg[i]+=2; } }
for(i=1;i<=v;i++) p[i]=i;
for(i=1;i<=v-1;i++) {
k = i;
for(j=i;j<=v-1;j++) if (deg[j+1]>deg[k]) k=j+1;
l = p[i];
p[i] = p[k];
p[k] = l;
m = deg[i];
deg[i] = deg[k];
deg[k] = m; }
pp = qq = 0;
System.out.println("Rank    Vertex    Degree\n");
for(i = 1;i<=v;i++) {
System.out.println(i+" "+p[i]+" "+deg[i]);
pp+ = deg[i];
qq+ = deg[i]*(deg[i] - 1); }
System.out.println();
rr = 10;
it = (deg[1] - deg[v])/(double)rr;
for(i=1;i<=10;i++) {
fr[i] = 0;
for(j=1;j<=v;j++)
if ((deg[j]>=(deg[v]+(i-1)*it)) & (deg[j]<(deg[v]+i*it)))
fr[i]++; }
System.out.println("Frequency distribution of degrees:");
for(i=1;i<=10;i++)
System.out.print(deg[v]+it/2.0+(i-1)*it+" ");
System.out.println();
for(i=1;i<=10;i++)
System.out.print(fr[i]/v+" ");
System.out.println("\ n");
mean = pp/v;
ss = 0;
for(i=1;i<=v;i++)
ss+ = Math.pow(deg[i] - mean,2);
var = ss/(v-1);
```

```java
skew=v/((v-2)*Math.sqrt(var));
System.out.println("Skewness of degree distribution: "+skew
+"\n");
h=v*qq/(pp*(pp - 1));
System.out.println("Aggregation index of the network: "+h);
if (h<=1) System.out.println("It is a random network.\n");
if (h>1) System.out.println("It is a complex network.\n");
h = var/mean;
System.out.println("Variation coefficient H of the network:
"+h);
System.out.println("Entropy E of the network: "+(var- mean));
if (h<=1) System.out.println("It is a random network.\n");
if (h>1) System.out.println("It is a complex network.\n");
ty=1;  //Binomial distri., pr = Crn pr qn-r, r=0,1,2,…, n;
ss = 0;
for(i=0;i<=rr - 1;i++) ss+=i*fr[i+1];
pp = ss/(v*(rr - 1));
qq = 1-pp;
pr[0] = Math.pow(qq,rr-1);
for(i-1;i<-rr-1;i++) pr[i]=(rr-i)*pp*pr[i-1]/(i*qq);
chi=xsquare(v, rr, pr, fr);
System.out.println("Binomial distribution Chi-square="+chi);
System.out.println("Binomial p="+pp);
k1 = 20.09;
coincidence(ty, k1, chi);
ty = 2;
//Poisson distri., pr = e-λλ r/r!, r = 0,1,2,…
pr[0]=Math.exp(- mean);
for(r=1;r<=rr - 1;r++) pr[r]=mean/r*pr[r-1];
chi=xsquare(v, rr, pr, fr);
System.out.println("Poisson distribution chi-square="+chi);
System.out.println("Poisson lamda="+mean);
k1 = 20.09;
coincidence(ty, k1, chi);
ty = 3;    //Exponential distri., F(x)=1-e-λ x, x ≥ 0
chi = 0;
for(i=1;i<=10;i++) {
k1 = deg[v]+it/2.0+(i-1)*it;
k2 = deg[v]+it/2.0+i*it;
pp = v*(Math.exp(-k1/mean) - Math.exp(-k2/mean));
chi+=Math.pow(fr[i] - pp,2)/pp; }
System.out.println("Exponential                    distribution
lamda="+1.0/mean);
k1 = 20.09;
coincidence(ty, k1, chi);
powerDistr(v, deg); }
```

```
public static double xsquare(int v, int rr, double p[],
double h[]) {
double hk,ss=0;
for(int i=0;i<=rr - 1;i++) {
hk=p[i]*v;
if (p[i]==0) hk=h[i+1];
ss+=Math.pow(p[i]*v - h[i+1],2)/hk;}
return ss; }
public static void coincidence(int ty, double k1, double ss){
if (ss<=k1)
if (ss>=0) {
if (ty==1) System.out.println("Degrees    are    binomially
distributed.\n");
if (ty==2) System.out.println("Degrees    are    Poisson
distributed.\n");
if (ty==3) System.out.println("Degrees are exponentially
distributed.\n");
if ((ty==1)|(ty==2)) System.out.println("It is a random
network"); }
if ((ss>k1) & ((ty==1)|(ty==2))) System.out.println("It is
likely not a random network\n");
if ((ss>k1) & (ty==3)) System.out.println("It is not an
exponential network\n"); }
public static void powerDistr(int v, int x[]) {
//Power law distri., f(x) = x - α, x ≥ xmin
int i, j, k, n, r, xmin;
double xmax,a,alpha,dd,maa;
int xminn[] = new int[v+1];
int xmins[] = new int[v+1];
double z[] = new double[v+1];
double zz[] = new double[v+1];
double cx[] = new double[v+1];
double cf[] = new double[v+1];
double dat[] = new double[10000];
k = 1;
xminn[1] = x[1];
for(i=1;i<=v;i++) {
n = 0;
for(j=1;j<=k;j++)
if (x[i]!=xminn[j]) n++;
if (n==k) {
k++;
xminn[k] = x[i]; } }
for(i=1;i<=k - 1;i++) xmins[i] = xminn[i];
for(i=1;i<=v - 1;i++) {
k = i;
```

```
for(j=i;j<=v - 1;j++)
if (x[j+1]>x[k]) k=j+1;
r = x[i];
x[i] = x[k];
x[k] = r; }
for(i=1;i<=v;i++) z[i]=x[v - i+1];
for(r=1;r<=v;r++) {
xmin = xmins[r];
n = 0;
for(i=1;i<=v;i++)
if (z[i]>=xmin) {
n++;
zz[n] = z[i]; }
maa = 0;
for(i=1;i<=n;i++) maa+=Math.log(zz[i]/xmin);
a=n/maa;
for(i=0;i<=n-1;i++) cx[i+1]=i*1.0/n;
for(i=1;i<=n;i++) cf[i]=1-Math.pow(xmin/zz[i],a);
dat[r] = 0;
for(i=1;i<=n;i++) {
cf[i] = Math.abs(cf[i]-cx[i]);
if (cf[i]>dat[r]) dat[r]=cf[i]; } }
dd = 1e+100;
for(i=1;i<=v;i++)
if (dat[i]<dd) dd=dat[i];
for(i=1;i<=v;i++)
if (dat[i]<=dd) {
k = i;
break; }
xmin = xmins[k];
n = 0;
for(i=1;i<=v;i++)
if (x[i]>=xmin) {
n++;
zz[n] = x[i]; }
maa = 0;
for(i=1;i<=n;i++) maa+=Math.log(zz[i]/xmin);
alpha = 1+n/maa;
alpha = (n - 1)*alpha/n+1.0/n;
System.out.println("Power law distribution KS D value="+dd);
if (dd<(1.63/Math.sqrt(n))) System.out.println("Degrees are
power law distributed, it is a scale-free complex network");
System.out.println("Power law alpha="+alpha);
System.out.println("Power law xmin="+xmin); }
}
```

2.   Network Analysis

Network research is based on graph theory. Network is represented by graph, i.e., vertices set and between-vertex relationship (edge) set. Between-vertex relationship is generally a binary relationship. Methodology of network analysis provides a formalism mechanism to represent, to measure and to simulate relational structures (Butts, 2009). To improve the binary relationship, weighting edges and representing multi-edges relationship with super-edge are allowed (Wasserman and Faust, 1994). Moreover, the dynamic changes of the relationship may be treated as a time series (Newman *et al.*, 2006).

Almost all of the algorithms and methods described in the previous chapters are methodology of network analysis.

Network topology shows universality. Various actual networks, regardless of their age, function and scope, all converge to a similar structure. Topological structure of network changes with the addition of vertices and edges. In other words, to explain the topological structure of a system, we first need to describe how the network came into being. For example, it is known that network topology is almost constant when some vertices are randomly removed from the network (Albert *et al.*, 2000).

A network will not evolve in a continuous way during its formation. A small change in control variables would result in a sudden and substantial structure changes (phase transitions) of network-for example, sudden changes in connection structure-which is called "explosive diffusion" (Achlioptas *et al.*, 2009; Bohman, 2009). A network will experience several phase transitions during its formation, in which the size of the largest (giant) component will change. Here the component refers to a set of vertices, in which any of the pairs of vertices are connected with edges. Suppose there are m randomly connected edges. The network may have many small components if $m < n/2$. If m is slightly larger than $n/2$, then there will be a unique component, which contains many vertices, while other components are still very small. With the addition of new edges the giant component will gradually increase, and finally cover all vertices (Bollobas, 2001). Occurrence and evolution of giant component is an important part of network dynamics.

The size of vertex set may significantly affect the resulting network. In order to avoid misleading conclusions, the definition of the set of vertices should include all the different entities, and these entities are relevant to the relationship being studied (Butts, 2009).

The classification of entities in network analysis is very important (Butts, 2009). For example, an ecologist wants to examine between-animal interactions. To construct a network, he (she) will take samples in some designated area, record all the interactions, and may also record the environment-animal relationships. In this situation, the classification of the feeding environments is necessary (Schoenly and Zhang, 1999a). This classification will greatly affect the network structure, thereby affecting the robustness of a given network.

ॐ CHAPTER 9 ॐ

Ecological Network Analysis: Research Advances

1. General Theory

In the past decade, network analysis became the mainstream of ecological theory. Many ecological problems, such as fundamental structure of ecological networks, robustness of network to disturbance, co-evolution and co-extinction, community succession, etc., can be addressed with network analysis. Without a deep insight into the ecological network structure, it is impossible to assess network robustness under species extinction, habitat destruction, and other human activities (Bascompte, 2009).

To approach the relationship between structure and functionality of network, three issues must be addressed (Bascompte, 2009): (1) The model must consider how population dynamics influences network structure and vice versa. (2) Network analysis must consider various interaction types. Various types of interactions jointly determine the stability of a network (Bastolla *et al.*, 2009). (3) Species invasion and climate change, etc., should also be considered.

If the set of some species and the set of interactions are hyper-represented in the network, i.e., if between-vertex interactions are hyper-represented based on the equivalent random network (Milo *et al.*, 2002), then they can be referred to as mosaics. Mosaics are modules for assembling a network. Usually a small mosaic is a three-layer food chain (Bascompte, 2009; Fig. 1; e.g., Predator-prey-resource, or Omnivore-prey-resource, etc.). In a food web, the relative frequency of occurrence of a network mosaic may be treated as the importance of the mosaic. All mosaics change

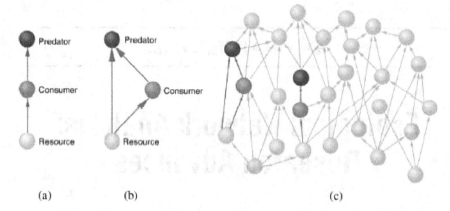

Figure 1.   Basic components of ecological networks (Bascompte, 2009): (a) Three-layer food chain modules; (b) Ominivory chain module; (c) The two modules can be treated as the mosaics in the network.

and thus become co-evolved geographical mosaics (Thompson, 2005). As a consequence, mosaics are important in the understanding of food web dynamics (Bascompte *et al.*, 2007; Stouffer *et al.*, 2007; Kondoh, 2008).

A large network is always organized from various small mosaics. The method for organizing mosaics to a large network will probably influence the stability of entire network, which is a theory proposed by May (1972). To search for evidence and facts for the theory is an active research area (Kondoh, 2008).

Ecosystem matter, energy and information networks are believed to be relevant. A full description of an ecological network should include all three networks that may contain the same compartments. However, information networks have not yet been studied to the same extent as networks of matter and energy cycling (Editorial, 2007). Research on the importance and descriptive power of information networks should be largely enhanced.

1.1. *Food webs*

Various food chains interact and connect to construct a network according to energy utilization relationships. The resulting network is a food web. The food web represents the trophic hierarchy and inter-relation of various species in an ecosystem. In the food web all species occupying the same trophic position make up a trophic level. For example, all plants in the food web constitute a trophic level called the first or "primary producers", all herbivores comprise the second or "primary consumer" trophic level, and all carnivorous animals constitute a third or "secondary consumer" trophic level. In addition, if there are more advanced carnivores that eat other carnivores, they will constitute an even higher trophic level. Through between-species trophic interaction in a community, the food web describes and quantifies the complexity of the ecosystem (May, 1972; Cohen, 1978; Pimm, 1982; Montoya *et al.*, 2006; Pascual *et al.*, 2006). Food webs are key topics of ecology (Bascompte, 2009).

In recent years, ecologists have analyzed the dynamics of model of large food webs. There were two components in the models. First they included the interaction network with dozens species which span several trophic levels (Pascual and Dunne, 2006; Montoya and Pimm, 2006); Second, they also considered practical body weight and metabolic rate (Brose and Williams, 2006). By including these factors in the models, the role of body weight ratio in preserving network's properties, and the influence of removing some species on species richness were studied (Berlow *et al.*, 2009). In general, the larger a food web, the more precise the conclusions.

Besides interaction strength and body weight ratio of predator to prey, mutualism (e.g., Plants-Pollinators, Plants-Seed Dispersers) is also considered in recent years' analysis of food web mosaics. These mutualistic food webs possess some characteristics: (1) Heterogeneity. Many species interact with a few species. Number of interactions of these few species is usually more than other species. (2) Nesting effect. (3) Between-species interactions are weak and asymmetric (Bascompte and Jordano, 2007). Bastolla *et al.* (2009) showed that species richness benefits from the mutualistic network structure. For a given number of interactions, the nest structure of mutualistic network will maximize the number of co-existed species.

A central question in food web theory is how static structures emerge from dynamics of interactive species (Cohen *et al.*, 1993). A possible way to address this question is to perform community assembly experiments. Most literature on dynamic assembly of multispecies ecosystems deals with competitive communities (Case, 1990; Morton *et al.*, 1996) or randomly wired ecosystems without trophic structure (May, 1973; Pimm, 1991).

The studies of food web properties have provided useful information for understanding ecosystem organization and its relationship with ecological stability (Pimm, 1991; Pimm *et al.*, 1991; Morin and Lawler, 1996; McCann, 2000). Most have centered on the scale-invariant (in terms of number of species) nature of food web patterns. For example, some network model predicted that food webs are robust to random species extinction but dependent upon some species with better connections. If these species are lost then the network will abruptly collapse (Dunne *et al.*, 2002; Memmott *et al.*, 2004; Montoya *et al.*, 2006).

It was found that deletion of the most connected species that are typical from the food webs with skewed degree distributions might result in coextinction of many other species by direct or indirect effects (Pimm, 1980; Dunne *et al.*, 2002; Montoya and Sole, 2002). Coextinction tends to occur in homologous species with similar genetic relationship (e.g., the same genera), which might result in a nonrandom pruning (Rezende *et al.*, 2007). Extinction of trophic levels occurs faster than coextinct species (Petchey *et al.*, 2008).

Most food webs have been determined by field sampling. However, sampling effort will exert a little effect on the topological properties (Martinez, 1991, 1994; Martinez *et al.*, 1999; Montoya and Sole, 2002, 2003). It was found that connectance is robust under different sampling efforts (Montoya and Sole, 2003). The same occurs for link distribution frequencies (Montoya and Sole, 2002).

1.2. *Ecoexergy*

Jorgensen and Fath (2006) concluded that some network changes are important for the power and ecoexergy of the network, but some changes that only remove energy flows from one place to another are not important. Changes that increase flows always increase the power and the ecoexergy. Natural selection of ecological networks should be further studied using various methods (Park *et al.*, 2006).

Some advances achieved in the past years include (Editorial, 2007): (1) increased network input results in a proportional increase of ecoexergy and power with unchanged network structure; (2) additional network links will only affect the power and ecoexergy if it gives additional exergy or energy transfer; (3) length increase of the food chain has positive effect on the power and ecoexergy of the network; (4) reduction of loss of exergy to the environment by respiration or as detritus yields a higher power and ecoexergy of the network; (5) faster cycling, through either faster detritus decomposed or increased transfer rates between two tropic levels, yields higher power and exergy; (6) input of additional exergy or energy cycling flows has greater effect than earlier in the food chain and the addition takes place.

1.3. Weak interactions

Montoya and Sole (2003) found that most interactions in complex food webs are weak, which confirmed some previous theoretical and empirical findings (Paine, 1992; Raffaelli and Hall, 1996; Berlow *et al.*, 1999; McCann, 2000). This is important for community stability and species coexistence (May, 1973; Laska and Wooton, 1998, McCann *et al.*, 1998; McCann, 2000; Zhang, 2011a) because: (1) weak interactions generate negative covariances between resources (prey), which promotes community-level stability; (2) negative covariances ensure species that interact weakly and dampen the destabilizing potential of strong interactions (McCann, 2000).

1.4. Degree distribution

Suppose S is the number of species, L is the number of actual links, then connectance C is defined as L divided by tire maximum possible number of links S^2. Some analyses found that food webs with different size were a roughly constant value of C from 0.1 to 0.15 (Martinez, 1992; Warren, 1994). Sometimes C varies in food webs as S increases from 0.27 to 0.02 (Montoya and Sole, 2003). Some found the inverse hyperbolic relationship between S and C based on the EcoWeb database (Cohen and Briand, 1984; Cohen *et al.*, 1990).

According to Montoya and Sole (2003), C was scale-variant and related to S according to a power law $S = C^a$ with an exponent $a \approx -1/2$. The constant connectance hypothesis, i.e., $C \approx 0.14$ despite changes in S, as reported in some previous studies (Pimm *et al.*, 1991; Havens, 1992; Martinez, 1992) did not hold for the 12 food webs investigated. The number of actual links L increases with S in a different manner from that predicted by both the link-species scaling law (*LSSL*) and the constant connectance hypothesis (*CCH*). Assuming the simplest relationship between L and S, $L = aS^b$, the *LSSL* holds that b must be close to one and on average the number of links per species in a web is constant and scale invariant at roughly 2, i.e., $L \approx 2S$ (Cohen *et al.*, 1990; Martinez, 1992). However, the *CCH* holds that L increases approximately as the square of S with $a < 1$ (a is the connectance C) and therefore $L = CS^2$ ($C = 0.14$; Martinez, 1991, 1992). Other studies rejecting *LSSL* indicated values of $b \approx 1.5$ (Sugihara *et al.*, 1989; Schoenly *et al.*, 1991; Havens, 1992; Martinez, 1994).

Montoya and Sole (2003) examined differences in degree distributions across some complex food webs and found that larger food webs have skewed degree distributions that strongly depart from those expected from random wiring (also see Zhang (2011a)), whereas food webs with fewer species have more homogeneous link distribution frequencies. This topological property highlights the importance of the position of species within food webs for their stability. It was found that in a species-rich network the most species had very few connections (many specialists) and only a few species were highly connected (fewer generalist preys and predators) (Montoya and Sole, 2003; Zhang, 2011a). The number of connections fluctuated around mean linkage density (L/S) when species richness was low. It was found that simulated networks always yielded Poisson degree distributions, which was independent of S.

To exploit how degree distributions affect community responses under species removals is an important topic in food web theory (Sole and Montoya, 2001; Dunne *et al.*, 2002; Montoya and Sole, 2002). According to Montoya and Sole (2003), the shape of the network of trophic interactions is highly dependent upon the species richness (Yodzis, 1980; Sugihara *et al.*, 1989; Martinez, 1992, 1994; Murtaugh and Kollath, 1997; Martinez *et al.*, 1999). The topological property might result from assembly processes and can be partially reconstructed by multi-trophic assembly models where dynamics are dominated by between-species weak interaction strengths.

1.5.　*Network structure and system stability*

Pinnegar *et al.* (2005) used a detailed Ecopath with Ecosim (EwE) model to test the impact of food web aggregation and the removal of weak linkages. They found that aggregation of a 41-compartment food web to 27- and 16-compartment systems greatly affected system properties (e.g., connectance, system omnivory, and ascendancy) and influenced dynamic stability. Highly aggregated webs recovered more quickly following disturbances compared to the original disaggregated model.

Food webs with skewed degree distributions showed two behaviors (Montoya and Sole, 2003). They exhibited high homeostasis when species were removed at random from the community but were very fragile when removals targeted generalist or most connected species. Food webs with Poisson degree distributions were highly fragile to both types of removals (random or directed).

Increased S implies a more complex distribution where few species play key roles in community persistence. Therefore, stochastic environmental fluctuations might affect species-rich communities less, but human perturbations (Wilson, 1992) might have larger effects. This is based on structural stability and does not consider dynamic effects of species deletion.

Many theories or hypotheses were proposed to explain diverse patterns of food web connectance (Warren, 1994). The most developed theory is that there is a relationship between C and different types of ecosystem stability, although various or even converse conclusions are drawn. Some models suggest that lower connectance involve higher local (May, 1973; Pimm, 1991; Chen and Cohen, 2001) and global (Cohen *et al.*, 1990; Chen and Cohen, 2001) stability, that is, the system recovers faster after a disturbance. Conversely, another theory suggests that a food web with higher connectance has more numerous reassembly pathways and can thus recover faster from perturbation (Law and Blackford, 1992).

2. Interaction Web Database

Interaction Web Database (http://www.nceas.ucsb.edu/interactionweb/resources.html) (IWDB) is a comprehensive database in which the data of diverse food webs have been recorded. They can be used to develop, test and validate ecological network models.

Some features of the database are summarized in this section. To find more details, go to homepage of Interaction Web Database.

	A	B	C	D	E	F	G	H	I	J	K	L	M
1				plant_com_name	Holly	Yew	Ivy	Mistlet	Dog ro	Haws	Rowan	Whitet	Wild s
2				plant_ge	Ilex	Taxus	Hedera	Viscum	Rosa	Cratae₍	Sorbus	Sorbus	Sorbus
3				plant_sp	aquifol	baccati	helix	album	canina	spp.	aucupa	aria	tormir
4	bird_com_name	bird_ge	bird_sp	no.	1	2	3	4	5	6	7	8	9
5	Blackbird	Turdus	merula	1	612	480	867	4	603	1559	505	257	0
6	Song Thrush	Turdus	philomelc	2	57	510	379	2	7	113	22	36	0
7	Mistle Thrush	Turdus	viscivoru:	3	370	233	119	251	12	147	32	69	0
8	Fieldfare	Turdus	pilaris	4	84	0	110	0	431	990	0	3	0
9	Redwing	Turdus	iliacus	5	582	55	168	1	4	536	0	170	0
.0	Robin	Erithacus	rubecula	6	15	26	76	3	1	19	7	2	0
.1	Blackcap	Sylvia	atricapilla	7	45	4	126	1	2	0	12	3	0
.2	Starling	Sturnus	vulgaris	8	0	837	279	0	0	198	63	11	0
.3	Waxwing	Bombycil	garrulus	9	0	0	0	0	0	0	0	0	0
.4	Woodpigeon	Columba	palumbus	10	32	0	0	0	15	96	0	0	0
.5	Ring Ouzel	Turdus	torquatus	11	0	3	12	0	0	0	0	0	0
.6	Garden Warbler	Sylvia	borin	12	0	0	0	0	0	0	1	0	0
.7	Lesser Whitethroat	Sylvia	curruca	13	0	0	0	0	0	0	3	0	0
.8	Magpie			14	0	0	0	0	0	0	3	27	0
.9	Jay			15	0	0	0	0	0	0	1	7	0
:0	Carrion Crow	Corvus	corone	16	0	0	0	0	0	0	0	24	0
:1	Common Whitethroat	Sylvia	communi	17	0	0	0	0	0	0	0	0	0
:2	Moorhen	Gallinula	chloropus	18	0	0	0	0	0	0	0	0	0
:3	Spotted Flycatcher	Musciapa	striata	19	0	0	0	0	0	0	0	0	0
:4	Green Woodpecker	Picus	viridis	20	0	0	0	0	0	0	0	0	0

Figure 2.   The data table of a food web in IWDB.

2.1. *Data types of food webs*

The database table of a food web in IWDB is as shown in Fig. 2.

(1) Main properties of database tables

Main properties of a database table include, for example:

(A) bird_com_name: common names.
(B) bird_ge: genus.
(C) bird_sp: species.

(2) Data Remarks

Every data are provided with source and data type explanation, for example (Fig. 3):

General information

In this paper, the authors examined the feeding patterns of grasshoppers from two arid grassland communities in Trans-Pecos, Texas. The studies took place from May until November in 1974 and 1975.

Ecological Network Analysis: Research Advances

A	B	C	D	E	F	G	H	I	J	K	L	M	N	O	P
Ant x Plant association matrix from Blüthgen et al. (2004) *Oikos* 106: 344-358. Counts of ant colonies attending extrafloral (EFN) and floral nectaries (FN) are provided. Independent abundance estimates for ant colonies based on sugar bait experiment, see Blüthgen & Fiedler (2004) *J. Anim. Ecol.* 73: 155-166. Ants with abundance = 0 were not recorded during bait experiments. For plants, number of plant individuals on which any insect were recorded on nectaries given as abundance estimate. Plant life forms: **cl** = climber, **he** = herb, **sh** = shrub, **tr** = tree, **pa**			**Plant spp.**	Adenia heterophylla	Aleurites rockinghamensis	Archontophoenix alexandrae	Archontophoenix alexandrae	Ardisia pachyrrhachis	Ardisia pachyrrhachis	Cardwellia sublimis	Castanospermum australe	Clerodendrum tracyanum	Crotalaria sp.	Cryptocarya hypospodia	
			Family (科)	Passifloraceae	Eup	Arec	Arecaceae	Myrs	Myrs	Prot	Fab	Lam	Fab	Laur	La
			Nectaries	EFN (花外蜜)	EFN	EFN	FN (花腺蜜)	EFN	FN	EFN	EFN	EFN	EFN	FN	FN
			Life form	cl	tr	pa	pa	sh	sh	tr	tr	sh	hr	tr	tr
			Abundance ⇔	4	1	1	3	17	1	4	4	5	1	1	
Ant spp.	Subfamily (亚科)	⇩	Total	4	1	1	4	29	1	7	11	6	1	1	1
Anonychomyrma gilberti	Dolichoderinae	48	63	1	0	0	0	1	0	1	2	0	0	0	
Camponotus spp. 'nocturnal' (novae-hol	Formicinae	88	8	0	0	0	0	2	0	0	0	0	0	0	
Camponotus vitreus	Formicinae	21	23	0	0	0	0	0	0	0	1	2	0	0	
Camponotus sp.1 (macrocephalus gp.)	Formicinae	5	5	0	0	0	0	0	0	0	0	1	0	0	
Camponotus sp.6 (gasseri gp.)	Formicinae	6	1	0	0	0	0	0	0	0	0	0	0	0	
Crematogaster aff. fusca	Myrmicinae	40	81	0	0	0	0	6	1	0	2	1	0	0	
Crematogaster aff. pythia	Myrmicinae	19	42	0	0	0	0	1	0	2	0	0	0	0	
Crematogaster sp.3	Myrmicinae	4	6	0	0	0	0	0	0	0	0	0	0	0	
Echinopla australis	Formicinae	6	2	0	0	0	0	0	0	0	0	0	0	0	
Leptomyrmex unicolor	Dolichoderinae	8	16	0	0	1	0	0	0	1	0	0	0	0	

Figure 3. Data remarks.

Data type

The authors recorded the identities of insect and plant species and their interactions.

Data are presented as a binary interaction matrix, in which cells with a "1" indicate an interaction between a pair of species, and a "0" indicates no interaction.

2.2. Types of food webs

2.2.1. Plant-pollinator food webs

These food webs describe the relationship between plants and pollinators in a region. Pollinators include various bees, wasps, butterflies and birds, etc. (Table 1). Usually the data were recorded as 0 or positive integers, in which 0 indicates no relation and positive integer denotes visiting frequency.

2.2.2. Host-parasite food webs

The Host-parasite food webs describe the relationship between all or partial hosts and parasites (Table 2). Usually the data were recorded as 0 or positive

Table 1.   A plant-pollinator food web (from IWDB).

pol_ge	pol_sp	no.	Aralia nudicaulis	Chimaphila umbellata	Clintonia borealis	Cornus caradensis	Cypripedium acaule	Linnaea borealis	Maianthemum canadense	Medeola virginiana
plant_ge / plant_sp	no.		1	2	3	4	5	6	7	8
Acmaeopsoides	rufula	1	0	0	0	1	0	0	0	0
Agiotes	stabilis	2	0	0	0	0	0	0	2	0
Ancistrocerus	sp.	3	0	0	0	2	0	0	0	0
Andrena	melanochroa	4	0	0	0	3	0	0	0	0
Andrena	miranda	5	0	0	0	2	0	0	1	0
Andrena	nivalis	6	0	0	0	1	0	0	1	0
Andrena	rufosignata	7	0	0	0	2	0	0	0	0
Andrena	sigmundi	8	0	0	0	1	0	0	0	0
Andrena	thaspii	9	0	0	0	1	0	0	0	0
Andrena	wheeleri	10	1	0	0	0	0	0	0	0
Andrena	wilkella	11	0	0	1	0	0	0	0	0
Andrena	w-scripta	12	1	0	0	7	0	0	0	0
Anthaxia	expansa	13	0	0	0	3	0	0	0	0
Anthonomus	sp.	14	0	0	0	1	0	0	0	0

Table 2. A host-parasite food web (from IWDB).

Host species	Sample size	Parasite genus Parasite species No.	Discocotyle salmonis 1	Octomacrum lanceatum 2	Triganodistomum simeri 3	Triganodistomum attenuatum 4
Osmerus mordax	50	1	0	0	0	0
Coregonus artedi	79	2	11.4	0	0	0
Coregonus hoyi	47	3	10.6	0	0	0
Coregonus clupeaformis	99	4	0	0	0	0
Prosopium cylindraceum	22	5	4.5	0	0	0
Oncorhynchus mykiss	15	6	0	0	0	0
Salvelinus fontinalis	13	7	0	0	0	0
Salvelinus namaycush	18	8	0	0	0	0
Catostomus commersoni	152	9	0	2.6	0	2.6
Catostomus catostomus	39	10	0	0	15.4	0

values, in which 0 indicates no relation, positive value denotes the number of parasites per host.

2.2.3. *Plant-ant food webs*

Plant-Ant food webs describe the relationship between ants and plants, usually expressed as the number of ants attracted by nectar (Table 3). Usually used data type is 0-positive integer type, in which 0 indicates no relation, positive integer denotes the number of ants per individual plant.

2.2.4. *Plant-seed disperser food webs*

Plant-seed disperser food webs describe the relationship between plants and seed dispersers (usually birds) (Table 4). Usually the data were recorded as 0 or positive integers, in which 0 indicates no relation, positive integer denotes the number of visits in a given time.

2.2.5. *Plant-herbivore and predator-prey food webs*

In Plant-herbivore food webs, data are presented as a binary interaction matrix, in which cells with a "1" indicate a trophic interaction between a pair of species and a "0" indicates no interaction.

See original database for description of Predator-prey food webs.

3. Topological Analysis of Some IWDB Food Webs

As discussed in preceding sections, a food web is a network to describe between-species trophic relationships. It also represents how the energy and materials flow through species. In the food web, the interacted species are connected by lines and arrows (i.e., links), and a species in the graph is a node (i.e., vertex). To study food webs helps to further understand the patterns of ecosystem organization and their relationship with ecological stability (Pimm, 1991; Pimm *et al.*, 1991; Warren, 1994; Morin and Lawler,

Table 3. A plant-ant food web (from IWDB).

Ant spp.	Subfamily	Abundance	Total	Passifloraceae EFN cl 4 4	Euphorbiaceae EFN tr 1 1	Arecaceae EFN pa 1 1	Arecaceae FN pa 3 4	Myrsinaceae EFN sh 17 29
		Family Nectaries Life form Abundance	Total					
Anonychomyrma gilberti	Dolichoderinae	48	63	1	0	0	0	1
Camponotus spp. 'nocturnal' (novae-hollandiae gp.,)	Formicinae	88	8	0	0	0	0	2
Camponotus vitreus	Formicinae	21	23	0	0	0	0	0
Camponotus sp.1 (macrocephalus gp.)	Formicinae	5	5	0	0	0	0	0
Camponotus sp.6 (gasseri gp.)	Formicinae	6	1	0	0	0	0	0

Table 4.   A plant-seed disperser food web (from IWDB).

Bird_com_name bird_ge bird_sp	plant_com_ name plant_ge plant_sp no.	Aglaia 1	Aporusa 2	Canthium 3	Chiso- cheton 4	Cissus hypoglauca 5
Crinkle-collared Manucode	1	0	0	0	1	0
Trumpet Manucode	2	0	0	0	7	13
Magnificent Riflebird	3	0	0	0	16	1
Blackbilled Sicklebill	4	0	0	0	5	1
Superb Bird of Paradise	5	0	1	0	23	1
Lawes' Six-wired Bird of Paradise	6	1	0	0	10	2
Magnificent Bird of Paradise	7	0	0	1	13	4
Raggiana Bird of Paradise	8	3	0	0	13	5
Blue Bird of Paradise	9	0	0	0	3	3

1996; McCann, 2000). However, many of these results look like non-natural laws because the data used is incomplete and error will be produced (Polis, 1991; Cohen *et al.*, 1993; Winemiller *et al.*, 2001).

The basic properties of the food web, including the actual number of links L, connectance C and their relationships should be fully taken into account in the food web study (Sugihara *et al.*, 1989). So far a few of the studies address between-species trophic links, degree distribution (i.e., hierarchical distribution; Bollobás, 1985), etc. These topological properties stress the importance of species in the stability of food web, which consider species' roles as both producer (incoming link) and consumer (outgoing link). Removal of the prominent species, which have most links to other species, will lead to direct or indirect effects on other species (Pimm, 1980; Solé and Montoya, 2001; Dunne *et al.*, 2002; Montoya and Solé, 2002).

Early studies on food webs began with MacArther's works. The main works during the period are: (1) food webs were in text and graphically expressed; (2) spatial uniformity, relationship linearity, and abstract between-species trophic relationships were assumed to study the stability and equilibrium of food webs. The food web studies during 1990s to 2000s focused on the general principles of link distributions. How to find general and stable patterns from food webs was one of the focuses in those studies (Cohen *et al.*, 1993). Most of the studies on community assembly have been based on between-species competition and stochastic linear assembly principles (May, 1983; Case, 1990; Morton *et al.*, 1996). The most recent studies on trophic networks are exploiting how between-species relationships affect the dynamics and stability of ecosystem (Navia *et al.*, 2010).

Through topological analyses on two food webs, predator-prey and parasitoid-host networks, Pimm *et al.* (1991) found the general model of the food webs. However, the conclusions drawn from parasitoids or predators may not fully represent the truth of typical parasites' role in the food webs. Unlike predators, parasites are very efficient in the food web's flow of energy and matter. The energy and matter flow of the large numbers of parasites from a host will profoundly affect the patterns and dynamics of the food web (Lafferty *et al.*, 2006b).

Recent studies have found that parasites can profoundly affect food web properties, such as nestedness, chain length and link density. Further, although most of the food web studies show that the vulnerability at the highest trophic level is the smallest, but if the parasites are included, then the species at the intermediate trophic level, rather than at the lowest trophic level, will have the highest vulnerability to natural enemies' attack. These results indicate that the food web which does not contain parasites is incomplete. Parasitic links are so important to ecosystem stability because they can increase the links and nestedness (Lafferty *et al.*, 2006a).

It is obvious that the topological analysis of Pimm *et al.* was not enough to draw a perfect reliable food web model. In this study we tried to consider parasitism in the topological analysis in order to provide a basis for further and more complete model development (Kuang and Zhang, 2011).

3.1.   *Materials and methods*

3.1.1.   *Materials*

3.1.1.1.   Data source

The data for topological analysis of food webs was derived from the food web studies of Lafferty *et al.*, conducted in Carpinteria Salt Marsh (CSM), California (Lafferty *et al.*, 2006a, b). The purpose of their study was to investigate the effects of parasites on the food web topology (Interaction Web Database: http://www.nceas.ucsb.edu/interactionweb/resources.html).

3.1.1.2.   Data description

CSM food web included four sub-webs. It is made of four sub-webs expressed as matrices. Four sub-webs are in the clockwise direction: Predator-prey sub-web, parasite-host sub-web, predator-parasite sub-web, and parasite-parasite sub-web. In the predator-parasite sub-web, a predator-parasite link was determined if a predator eats a prey who has been parasitized by parasite(s). Parasite-parasite sub-web includes hyperparasites (Kuris, 1990; Lafferty *et al.*, 1994; Huspeni and Lafferty, 2004). Six trophic levels are included in the predator-prey sub-web.

3.1.1.3.   Data conversion

Before the analysis, species were labeled by ID codes (the following table). After conversion, the data/data editors/matrix editor in the UCINET software was opened and then coded data was pasted. Using Matrix Editor, files were saved in ".##h" format. Finally, File/Open/Ucinet dataset/network option was used in Netdraw software to select and open the ".##h" file, and then saved in ".net" format by File/Save data as/Pajek/Net file. The resultant four ".net" files formed the basis for topological analysis using Pajek.

1	Marine detritus	33	*Macoma nasuta*	65	Bonaparte's Gull	97	Eugregarine
2	Terrestrial detritus	34	Protothaca	66	Long-billed Curlew	98	Plasmodium
3	Carrion	35	*Tagelus spp.*	67	Surf Scoter	99	Nematode in tagelus
4	Macroalgae	36	Cryptomya	68	Bufflehead	100	*Spirocamellanus perarai*
5	Epipellic flora	37	*Mytilus galloprovincialis*	69	Clapper rail	101	*Baylisascaris procyonis*
6	Emergent vascular plants	38	Geonemertes	70	Cooper's Hawk	102	Acanthocephalan in Gillichthys
7	Sumergent vascular	39	American Coot	71	Northern Harrier	103	*Euhaplorchis californiensis*
8	Phytoplankton	40	Mallard	72	*Leptocettus armatus*	104	*Himasthla rhigedana*
9	Oligochaete	41	Killdeer	73	*Gillycthys mirabilis*	105	*Probolocoryphe uca*
10	Capitella capitata	42	Green-winged teal	74	*Urolopaus halleri*	106	*Himasthla species B*
11	Phoronid	43	*Cleavlandia ios*	75	Procyon locator	107	*Renicola buchanani*
12	Spionidae	44	Semipalmated Plover	76	Great Blue Heron	108	Acanthoparyphium sp.
13	Eteone lightii	45	Greater Yellowlegs	77	Snowy Egret	109	Cataropis johnstoni
14	Turkey Vulture	46	*Hemigrapsus oregonensis*	78	Black-crowned Night heron	110	Large xiphideocercaria
15	*Corophium sp*	47	*Fundulus parvipinnis*	79	Double-Crested Cormorant	111	*Parorchis acanthus*

(Continued)

(Continued)

No.	Name	No.	Name	No.	Name
16	Harpacticoid	48	Western Sandpiper	80	Great Egret
17	Ostracods	49	Dunlin	81	Pied Billed Grebe
18	Anisogammarus confervicolus	50	Least Sandpiper	82	Osprey
19	Traskorchestia	51	Forster's Tern	83	Triakis semifasciata
20	Uca crenulata	52	Dowitcher	84	Portunion conformis
21	Neotrypaea	53	Green Heron	85	Picornavirus
22	Upogebia	54	Belted Kingfisher	86	Nerocila californica
23	Atherinops affinis	55	American Avocet	87	Orthione
24	Mugil cephalus	56	Pachygrapsus crassipes	88	Ergasilus auritious
25	Cerithidea californica	57	Willet	89	Aedes taeniorhynchus
26	Acteocina inculcata	58	Black-bellied Plover	90	Culex tarsalis
27	Melampus	59	California Gull	91	Leech (glossiphonidae)
28	Assiminea	60	Whimbrel	92	Proleptus
29	Trichocorixia	61	Mew Gull	93	Carcinonemertes
30	Ephydra larva	62	Marbled Godwit	94	Gyrodactylus
31	Mosquito larva	63	Ring-billed gull	95	Trichodina
32	Ephydra adult	64	Western Gull	96	Eugregarine

112	Austrobiharzia
113	Cloacitrema michiganensis
114	Phociremoides ovale
115	Renicola cerithidicola
116	Small Cyathocotylid
117	Stictodora hancocki
118	Mesostephanus appendiculatoides
119	Pygidiopsoides spindalis
120	Microphallid 1
121	Hysterolecitha
122	Parvatrema
123	Microphallid 2
124	Galactosomum
125	Tetraphyllidean
126	Tetraphyllid fish
127	Trypanorynch
128	Dilepidid

3.1.2.   *Methods*

3.1.2.1.   Pajek software

Pajek is the software to analyze large and complex networks. It is a fast and visualized program. It is unique to calculate the networks with millions of nodes. It is mainly used to conduct global analysis on complex networks.

3.1.2.2.   Some properties of food webs

(1) Classification of species

   Species were classified into three categories, top (trophic) species, intermediate (trophic) species and base (trophic) species (Pimm *et al.*, 1991).

(2) Degree

   Degree is the most basic property for a complex network. The degree of a node is defined as the number of its connected nodes. In general, the more the degree of a node, the more important the node is. In an oriented network, the degree is the sum of incoming degree and outgoing degree. Using the In/Out/All commands of Net/Partitions/Degree menu of Pajek, the degree, incoming degree and outgoing degree can be calculated. The proportions of three categories of species can be obtained by calculating degrees of these species.

(3) Chain cycle

   Chain cycle is a closed loop in the food chain. Cannibalism is a kind of chain cycle. Chain cycle can be obtained by using Net/Count/4-rings/directed/cyclic in Pajek.

(4) Connectance and link density

   Connectance is the ratio of realized trophic interactions to possible interactions. In the calculation of connectance, the number of possible interactions is S^2 if cannibalism is considered, or else it is $S(S-1)$. Link density is equal to the ratio of total number of links to the total number of species.

(5) Chain length

   Chain length refers to the number of links of the path between the base species and the top species through the chain of two adjacent species. The chain length or between-species distances can be calculated by Net/k-neighbors/output in Pajek.

(6) Omnivorous species

An omnivorous species is dependent upon more than one trophic levels. Omnivorous species make the boundaries between trophic levels blurred. Omnivory is the ratio of the number of closed omnivorous links to the number of top species (Sprules and Bowerman, 1988). A closed omnivorous link refers to that a predator feeds on the two preys with different trophic levels along the same food chain.

3.2. Results

3.2.1. Species analysis

The results of species analysis on four sub-webs are indicated in Table 5. Pimm *et al.* (1991) pointed out that the proportions of top species, intermediate species and base species are generally constants. According to our results, however, the number of base species is kept constant but the number of top species declines remarkably (from 33 species to three species) when parasites are added. Thus the proportions change sharply (Fig. 4).

3.2.2. Cycle analysis

There is not a single cycle in the predator-prey, predator-parasite, and parasite-host sub-webs. Contrarily there are 508 cycles in the parasite-parasite sub-web, and there are 85214 cycles in the complete food web. In the studies of Pimm *et al.* (1991), however, rare cycles appeared for food webs without parasites.

3.2.3. Link analysis

There are 992 links in the predator-prey sub-web. The connectance and link density is 0.29 and 11.95 respectively (Fig. 5).

Figure 5 shows that the species *Pachygrapsus crassipes* and *Hemigrapsus oregonensis*, with 45 and 43 links respectively, are the two most significant species in the predator-prey sub-web. Second by *Fundulus parvipinnis* (35 links). Turkey vulture has only one link.

There are 1,260 links in the predator-parasite sub-web. The connectance and link density of this sub-web are 0.16 and 9.84 respectively (Fig. 6).

Table 5.   Species analysis of complete food web.

(Sub-)Food web	Category	Number	Total No. Species	Percent (%)	Species ID code
Predator-Prey	T	33	83	39.76	14,42,44,45,51,53–55,57–71,74–83
	I	42		50.60	9–13,15–41,43,46–50,52,56,72,73
	B	8		9.64	1,2,3,4,5,6,7,8
	O	0		0	—
Predator-Parasite	T	44	128	34.38	84,86–128
	I	0		0	—
	B	63		49.22	10,12–14,16,18,20–26,28,33–36,39–83
	O	21		15.41	1–9,11,15,17,19,27,29–32,37,38,85
Parasite-Host	T	47	128	35.72	34,35,38–40,42–83
	I	0		0	—
	B	41		32.03	84–88,91–100,102–125,127,128
	O	40		31.25	1–33,36,37,41,89,90,101,126
Parasite-Parasite	T	2	45	4.44	85,98
	I	17		37.78	103–111,113–120
	B	2		4.44	84,90
	O	24		53.33	86–89,91–97,99–102,112,121–128
Complete food web	T	3	128	2.34	89,101,126
	I	117		91.41	9–88,90–100,102–125,127,128
	B	8		6.25	1,2,3,4,5,6,7,8
	O	0		0	—

Note: T-top species; I-intermediate species; B-base species; O-species outside web. There are no intermediate species in the predator-parasite and parasite-host sub-food webs due to the incomplete data.

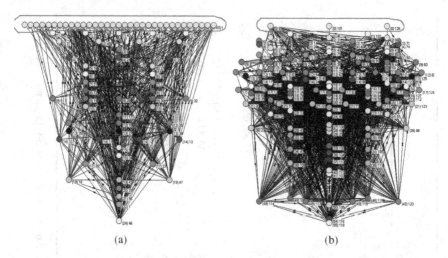

(a)   (b)

Figure 4.   Comparison of food webs with (b) and without (a) parasites. The number in parentheses is total links (degree, or incoming degree + outgoing degree) and the number outside parentheses is species ID code. From top to bottom layers the number of links of each species increases.

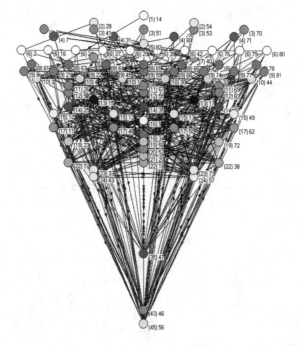

Figure 5.   Predator-prey sub-web.

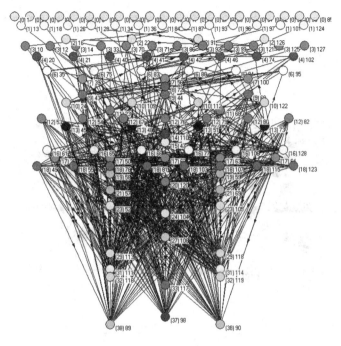

Figure 6.   Predator-parasite sub-web.

Figure 6 shows that the species *Aedes taeniorhynchus* and *Culex tarsalis*, with 38 links respectively, are the two most significant species in the predator-parasite sub-web, seconded by Plasmodium (37 links). Some species, such as marine detritus and Picornavirus, have no links. They are isolated species.

There are 1984 links in the parasite-host sub-web, and the connectance and link density of this web is 0.24 and 15.5 respectively (Fig. 7). *Himasthla rhigedana*, *Himasthla* species B, *Renicola buchanani*, and *Catatropis johnstoni* have the most links (40 links) in the parasite-host sub-web. Species, such as Killdeer, etc., have no links.

In total, 344 links are found in the parasite-parasite sub-web and the connectance and link density is 0.34 and 7.64 respectively (Fig. 8). In parasite-parasite sub-web, *Mesostephanus appendiculatoides* has the most links (27 links) and *Himasthla rhigedana* has the least links (16 links).

Totally there are 4,580 links in the complete food web. The connectance and link density for the food web is 0.56 and 35.78 respectively (Fig. 9).

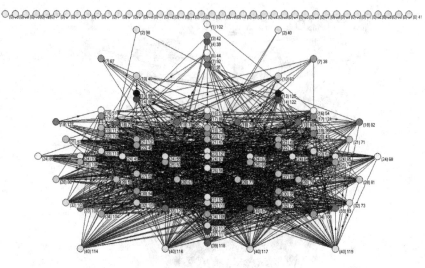

Figure 7.   Parasite-host sub-web.

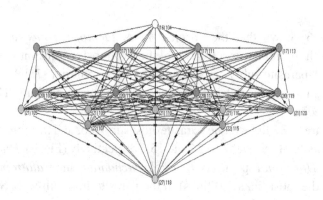

Figure 8.   Parasite-parasite sub-web.

In the complete food web, small cyathocotylid (93 links), *Stictodora hancocki* (93 links), *Mesostephanus appendiculatoides* (95 links), and *Pygidiopsoides spindalis* (92 links) are the most significant species. *Baylisascaris procyonis* has one link only.

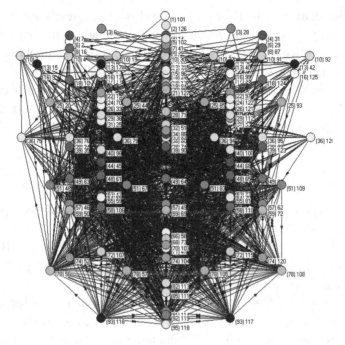

Figure 9.   Complete food web.

Table 6.   Parameters of web links.

(Sub-)Food web	Total links	Percent (%)	Connectance	Link density	Maxi. No. links	Total No. Species
Predator-Prey	992	21.66	0.29	11.95	45	83
Predator-Parasite	1260	27.51	0.16	9.84	38	128
Parasite-Host	1984	43.32	0.24	15.5	45	128
Parasite-Parasite	344	7.51	0.34	7.64	27	45
Complete food web	4580	100	0.56	35.78	95	128

From Table 6 we can find that the links of predator-prey sub-web accounts for only 21.66% of the total links of complete food web, while the links of parasite-host sub-web (43.32%) and predator-parasite sub-web (27.51%) account for 70.83% of the total. This result stresses the importance of parasitism in the food web.

The link density of predator-prey sub-web is 11.95, greatly less than the 35.78 of complete food web, which means that the addition of parasitism in the food web will remarkably increase link density. The number of top species, intermediate species and base species in the predator-prey sub-web is 275, 641 and 76, respectively, much different from the number of 41, 4463, and 76 in complete food web. We may find from these results that top species decline and intermediate species increase sharply after parasitism is added.

3.2.4. *Chain length*

For both predator-prey sub-web and complete food web, the base species are species with ID code 1 to 8.

Similar to the analysis on the No. 1 species, as indicated in Fig. 10, the K-neighbor/output analyses on No. 2 to No. 8 in the predator-prey sub-web is conducted, as shown in Table 7.

Pimm *et al.* (1991) pointed out that the chain length for top species is typically two or three, and one is relatively rare (led by incomplete information), and the chain length larger than three occurs seldom. The corresponding number of trophic levels is three or four. In present analysis there are six trophic levels and most chain lengths are three. The results are in accordant with the Pimm *et al.* (1991).

For the complete food web, most of the chain lengths are three and some are more than three (Fig. 11; No. 8 and No. 6 species).

As can be seen from Fig. 11, the chain lengths for the food web with parasitism are larger slightly than the web without parasitism.

3.2.5. *Analysis on omnivorous species*

There are many omnivorous species in the food web. In the predator-prey sub-web, the omnivory is increased as the trophic level rises (Lafferty *et al.*, 2006b; Table 8).

3.3. *Conclusions and discussion*

Most of the food webs so far lack parasitism. Actually, once parasitism is added in the network, the traditional top species would not still be at the highest trophic level because most of the species are parasitized by one or

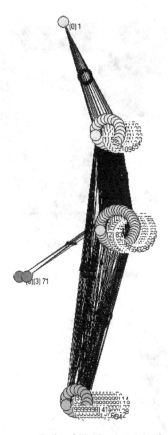

Figure 10. K-neighbor/output analysis of ID No. 1 species in the predator-prey sub-web. The species No. 9,999,998 means that it is not reachable to No. 1 species. Among species reachable to No. 1 species, the maximum chain length is three.

Table 7. Distribution of chain length for No. 1 to No. 8 species in the predator-prey sub-web.

(Sub-)Food web	1	2	3	4	5	6	7	8	Maximum Chain length
Predator-Prey	3	3	3	3	3	2	3	3	3
Complete food web	3	3	3	3	3	5	3	4	5

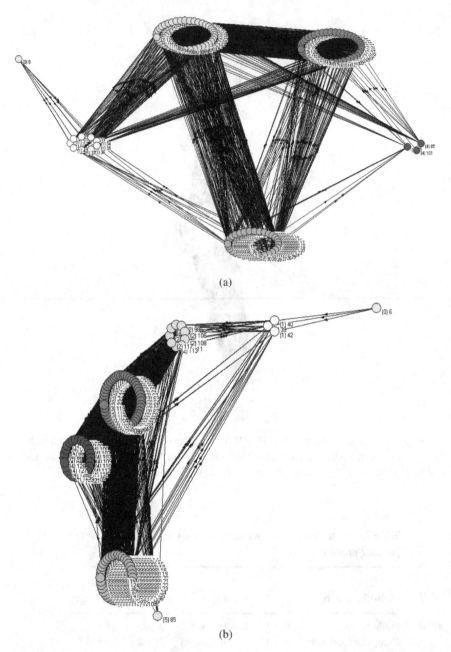

(a)

(b)

Figure 11.   K-neighbor/output analysis of ID No. 8 (a) and No. 6 (b) species in the complete food web. Among species reachable to No. 8 (No. 6) species, the maximum chain length is four (5).

Table 8. Change of omnivory with trophic level.

Trophic level	1	2	3	4	5	6
Omnivory	1.0	1.5	2.5	3.0	3.6	4.4

more parasites (Polis, 1991). The addition of parasitism greatly increases the complexity of food webs and alters the properties of food webs. As indicated in the present study, the major changes include the following aspects:

(1) **Structural changes in species.** The proportion of top species, intermediate species and base species change after parasitism is added. The number of top species declines and the number of intermediate species increases sharply. The number of base species will not change on the addition of parasitism. If all parasite species are treated as top species, however, the proportion of top species will increase and the proportion of intermediate species and base species will decline (Huxham *et al.*, 1995).

(2) **Increase in chain cycles.** Rare chain cycles were found in the food web with predators and preys only (Pimm *et al.*, 1991). Different from the observation of Pimm *et al.* (1991), the between-parasite cycles increase largely once parasitism is added. Moreover, there will be more cycles between predators and preys due to the addition of parasites.

(3) **Increase in links.** If the parasitism is added, the number of links and link density will increase, and the proportions of top species, intermediate species and base species will be altered. In average, the links between parasites and hosts are much more than that between predators and preys (Lafferty *et al.*, 2006b). The links between predators and parasites are greater than the links between predators and preys due to the remarkable existence of parasites in hosts.

The links between parasites and hosts increase more than that of the total number of links, thus the link density also increases. Another study has proved that link density increases from 5.36 to 8.64 (Amundsen *et al.*, 2009).

A large numbers of parasites serve as both consumers and producers, thereby the number of intermediate species increases greatly, which results in the significant changes of the proportions of top species, intermediate species and base species.

(4) **A slight increase in the chain length.** According to Pimm *et al.* (1991), most of the chain lengths are two or three. Average chain length increases after parasitism is added (Thompson *et al.*, 2005), as proved in the present study.

(5) **Increase in omnivory.** Parasites can consume several trophic levels, thereby omnivorous species and omnivory increase (Huxham *et al.*, 1995). Some research proved that omnivory increases from 1.86 to 2.07 (Amundsen *et al.*, 2009).

Further research may center on the following aspects:

(1) This study was based on the food web data from Carpinteria Salt Marsh, and some other published data. However, to obtain complete results, we need more data and use some model as cascade model, etc., to validate conclusions or exploit mechanism. More interaction types, e.g., mutualism (Callaway, 1995; Bruno *et al.*, 2003; Bascompte and Jordano, 2007; Dormann, 2011), etc., should also be considered.

(2) Predator and prey overlap graphs are suggested to be developed to analyze topological holes for species with lower abundance.

(3) Dynamic analysis, such as agent-based modeling (Zhang, 2011b; Zhang, 2012; to be published), etc., is suggested to be used in the dynamic analysis of network structure.

4. Calculation of Degree Distribution and Detection of Network Type

Interaction Web Database (National Center for Ecological Analysis and Synthesis, 2011; http://www.nceas.ucsb.edu/interactionweb/) was chosen as the data source of the present study. Interaction Web Database contains seven food webs, namely Anemone-Fish, Host-Parasite, Plant-Ant, Plant-Herbivore, Plant-Pollinator, Plant-Seed disperser, and Predator-Prey sub-webs. For each web, the species with corresponding interspecific relationship but not all species in the ecosystem or community, were included. Each of seven food webs was used to calculate degree distribution and detect network type (Zhang and Zhan, 2011).

For Anemone-Fish web, we used the data of Fautin and Allen (1997) and Ollerton *et al.* (2007), as indicated in Table 3. The data for other webs were chosen as follows:

Host-Parasite webs: We used the data for Canadian freshwater fish and their parasites (Arthur *et al.*, 1976), which were from the investigation to seven water systems. Moreover, the data from Cold Lake (Leong *et al.*, 1981; 10 hosts and 40 parasites) and Parsnip River (Arai *et al.*, 1983; 17 hosts and 53 parasites) were also used.

Plant-Ant web: The data of Bluthgen (2004) from tropical rain forests, Australia, were used. There ware 51 plants and 41 ants in this web.

Plant-Pollinator webs: We used a set of data collected from KwaZulu-Natal, South Africa (Ollerton *et al.*, 2003; 9 plants and 56 pollinators), and the data from Canada (Small, 1976; 13 plants and 34 pollinators).

Plant-Herbivore web: The data from Texas, USA (Joern, 1979; 54 plants and 24 herbivores) were used.

Predator-Prey webs: Four sets of data (Berwick, Catlins, Coweeta and Venlaw) were used. The major species included algae, fish, arthropods and amphibians.

Plant-Seed disperser webs: Two sets of data were used. One from a forest in Papua New Guinea (Beehler, 1983; 31 plants and 9 birds), and one from a tropical forest in Panama (Poulin *et al.*, 1999; 13 plants and 11 birds).

A typical raw data in Interaction Web Database is indicated in Table 9.

Table 9.   An example of the data of Interaction Web Database.

Species	Unidentified detritus	Terrestrial invertebrates	Plant material	*Achnanthes lanceolata*
Unidentified detritus	0	0	0	0
Terrestrial invertebrates	0	0	1	0
Plant material	0	1	0	0
Achnanthes lanceolata	0	0	0	0

General information In this paper, the authors examined the feeding patterns of grasshoppers from two arid grassland communities in Trans-Pecos, Texas. The studies took place from May until November in 1974 and 1975.

Date type The authors recorded the identities of insect and plant species and their interactions. Data are presented as a binary interaction matrix, in which cells with a "1" indicate an interaction between a pair of species, and a "0" indicates no interaction.

Table 10.   The data transformed from Table 9.

ID of Taxon 1	ID of Taxon 2	Value
1	2	1
1	3	1
2	3	1
2	4	1
3	4	1

In Table 9, the values 1 and 0 represent having or not having interspecific trophic relationships. The values neither 1 nor 0 represent frequencies and these values were transformed to 1 in present study. Table 9 should be transformed to the format needed by the Java algorithm above, as indicated in Table 10.

4.1. Results

The data of Anemone-Fish web is indicated in Table 11.

Table 11.   Species and ID of Anemone-Fish web.

Genera	Species	ID	Genera	Species	ID
Amphiprion spp.	Akallopisos	1	Amphiprion spp.	percula	19
Amphiprion spp.	Akindynos	2	Amphiprion spp.	perideraion	20
Amphiprion spp.	Allardi	3	Amphiprion spp.	polymnus	21
Amphiprion spp.	Bicinctus	4	Amphiprion spp.	rubrocinctus	22
Amphiprion spp.	chrysogaster	5	Amphiprion spp.	sandaracinos	23
Amphiprion spp.	chrysopterus	6	Amphiprion spp.	sebae	24
Amphiprion spp.	clarkii	7	Amphiprion spp.	tricinctus	25
Amphiprion spp.	ephippium	8	Premnas	biaculeatus	26
Amphiprion spp.	frenatus	9	Heteractis	crispa	27
Amphiprion spp.	fuscocaudatus	10	Entacmaea	quadricolor	28
Amphiprion spp.	latezonatus	11	Heteractis	magnifica	29
Amphiprion spp.	latifasciatus	12	Stichodactyla	mertensii	30
Amphiprion spp.	leucokranos	13	Heteractis	aurora	31
Amphiprion spp.	mccullochi	14	Stichodactyla	gigantea	32
Amphiprion spp.	melanopus	15	Stichodactyla	haddoni	33
Amphiprion spp.	nigripes	16	Macrodactyla	doreensis	34
Amphiprion spp.	ocellaris	17	Heteractus	malu	35
Amphiprion spp.	omanensis	18	Cryptodendrum	adhaesivum	36

Table 11 is transformed to the data type needed by the Java algorithm, asindicated in Table 12.

Output results of the Java algorithm for Anemone-Fish web are as follows:

Skewness of degree distribution: 0.2739383901373063
Aggregation index of the network: 1.519811320754717
It is a complex network.
Variation coefficient H of the network: 3.361428571428571
It is a complex network.
Binomial distribution Chi-square = 84779.33198479559
Binomial p = 0.2222222222222222
It is likely not a random network
Poisson distribution chi-square = 462.75941519476396
Poisson lamda = 4.444444444444445
It is likely not a random network
Exponential distribution lamda = 0.22499999999999998
Degrees are exponentially distributed.
Power law distribution KS D value = 0.0
Degrees are power law distributed, it is a scale-free complex network
Power law alpha = NaN
Power law xmin = 14

It is obvious that the food web is a complex network.

The results for all food webs are listed in Tables 13 and 14.

The results show that the degree distribution of most of the food webs is power law and exponential distribution, and all of the food webs are complex networks.

From variation coefficient and aggregation index in Tables 5 and 6, we can find that all values are greater than one and all webs are thus complex networks. Plant-pollinator web (Ollerton *et al.*, 2003) is the most complex, seconded by Predator-prey web (Catlins) and Plat-ant web (Bluthgen, 2004), the complexity of Plant-seed disperser web (Poulin, 1999) is the lowest. It can be found that the skewness of Plant-pollinator web (Ollerton *et al.*, 2003) is the smallest and its degree distribution is the most skewed, which reveals it is the most complex network.

Table 12.   A data type of anemone-fish web.

ID of taxon 1	ID of taxon 2	Value	ID of taxon 1	ID of taxon 2	Value
1	27	1	7	32	1
1	28	1	7	34	1
1	29	1	8	28	1
1	30	1	8	30	1
1	31	1	8	31	1
1	32	1	9	27	1
1	33	1	9	28	1
1	34	1	9	29	1
1	35	1	9	32	1
1	36	1	10	29	1
2	27	1	10	30	1
2	28	1	10	32	1
2	29	1	11	27	1
2	30	1	11	29	1
2	31	1	11	32	1
2	32	1	12	27	1
2	33	1	12	29	1
3	27	1	12	30	1
3	28	1	13	27	1
3	29	1	13	33	1
3	30	1	13	34	1
3	31	1	14	27	1
3	33	1	14	28	1
3	34	1	14	33	1
4	29	1	15	28	1
4	30	1	15	32	1
4	31	1	16	27	1
4	33	1	16	30	1
5	27	1	17	29	1
5	28	1	17	30	1
5	29	1	18	27	1
5	30	1	18	28	1
5	31	1	19	28	1
5	32	1	20	30	1
6	27	1	21	27	1
6	28	1	22	30	1
6	30	1	23	28	1
6	31	1	24	29	1
7	27	1	25	33	1
7	29	1	26	28	1

Table 13.   Summary of results for calculation of degree distribution and network type.

	Anemone-fish web	Host-parasite webs			Plant-ant web	Plant-pollinator webs	
Skewness of degree distribution	0.2739	0.2524	0.2822	0.2404	0.1626	0.2065	0.2330
Aggregation index of the network	1.5198	1.6913	1.7425	1.6709	1.8597	3.1461	1.3843
Variation coefficient of the network	3.3614	4.0615	3.7425	4.0627	6.3754	7.8742	3.3478
Binomial distribution Chi-square	84779.33	316459.4	1500517	53631.9	346819.3	156.63	536.45
Binomial p	0.2222	0.2099	0.1556	0.1905	0.1715	0.0325	0.2577
Poisson distribution Chi-square	462.759415	538.79	498.22	1661.19	11352.2	1007.7	1197.5
Poisson lamda	4.4444	4.3333	3.6400	4.5143	6.1957	3.1692	6
Exponential distribution lamda	0.2249	0.2308	0.2747	0.2215	0.1614	0.3155	0.1667
Power law distribution KS D value	0	0	0	0	0.1586	0	0
Power law alpha	—	—	—	—	—	—	—
Power law Xmin	14	15	15	17	6	35	18
Type of degree distribution	Exponential, power law	Power law	Power law	Power law	Power law	Power law	Power law
Network type	Complex network	Complex network	Complex network	Complex network	Complex network	Complex network	Complex network

Table 14.   Summary of results for calculation of degree distribution and network type.

	P-H web	Plant-seed disperser webs		Predator-prey webs			
Skewness of degree distribution	0.2261	0.1969	0.3612	0.1735	0.2069	0.2313	0.1918
Aggregation index of the network	1.8139	1.6254	1.2334	1.7743	2.0198	1.8211	1.7875
Variation coefficient H of the network	4.6468	4.8003	2.0656	5.7552	5.6528	4.6157	5.3199
Binomial distribution Chi-square	48066.1	1451.8	898.8	256.2	180662.6	382.9	3897.6
Binomial p	0.1182	0.2	0.3333	0.1252	0.1043	0.0958	0.1643
Poisson distribution chi-square	2167.1	1306.7	64.4	8483.6	945.7	1286.7	2241.8
Poisson lamda	4.4359	5.95	4.4167	6.0759	4.4898	4.3448	5.4203
Exponential distribution lamda	0.2254	0.1681	0.2264	0.1646	0.2227	0.2302	0.1845
Power law distribution KS D value	0	0	0	0	0	0	0
Power law alpha	—	2.9967	—	—	—	—	—
Power law Xmin	23	6	11	35	27	26	26
Type of degree distribution	Power law	Exponential, power law	Exponential, power law	Exponential, power law	Exponential, power law	Power law	Exponential, power law
Network type	Complex network	Complex network	Complex network	Complex network	Complex network	Complex network	Complex network

Note: P-H web means Plant-herbivore web.

The results show that the degree distribution of most of the food webs is power law and exponential distribution, and all of the food webs are complex networks.

4.2. Discussion

Different from classical distribution patterns (bionomial distribution, Poisson distribution, and power law distribution, etc.), both network type and network complexity can be calculated and compared using the indices above, i.e., aggregation index, coefficient of variation, skewness, etc. We suggest that they should be used in the network analysis.

Other indices to detect aggregation intensity can also be used in network analysis. For example, the Lloyd index:

$$L = 1 + (s^2 - \bar{u})/\bar{u}^2,$$

where $\bar{u}$, s^2 are the mean and variance of network degree respectively. The network is a random network, if $L \leq 1$. It is a complex network, if $L > 1$, and network complexity increases with L.

5. Brain Network

Nervous system is a complex network. Imagine that the nervous system with only 302 neurons can keep the worm *Caenorhabditis elegans* alive, sense the surroundings, make decisions and issue commands to the body (Zimmer, 2011). Human brain is a much complex network. There are 100 billion neurons, constituting a network with 100 trillion connections, in human brain. Human brain is not an ecological network. However its network structure and behaviors is the same as complex ecological systems (see Zimmer, 2011 for details). In recent years, human brain researches are focused on network analysis. Uncovering human brain's intricate network will provide critical clues to understand the origins of our thinking, mental and body disorders. So far various mathematical models were developed to analyze brain network. Some results on human brain network are thus described here.

A model by Sporns *et al.* (Sporns *et al.*, 2000; Sporns, 2010) used 1600 simulated neurons which arrayed around the surface of sphere. They linked each neuron to others. At any time each neuron has a tiny chance to spontaneously firing. Once a neuron fires, it has a tiny chance to trigger its adjacent neurons to fire as well. First, Sporns *et al.* connected each neuron to its adjacent neurons. The resulting network produced random and small flickers of activity. Second, they linked every neuron to every neurons in entire brain and the resulting network yielded thoroughly different pattern — the brain began to switch on and off in regular pulses (Zimmer, 2011). Third, every neuron is admitted to fire both its adjacent neurons and other long-distant neurons. The network became so complex. As neurons began to fire, they sparked great glowing patches of activity swirling across the brain, and some patches collided with one another and some traveled around the brain in circles (Zimmer, 2011). Sporns' experiment showed that network structure shapes its activity pattern.

Rockmore and Pauls, two mathematicians at Dartmouth College compared the brain and stock market, both of which consist of small units (traders, neurons). Traders may influence each other in buying and selling, and this might further result in entire network and make the stock market rise or fall (Zimmer, 2011). The entire network can influence the lower levels. For example, as the rise or fall of stock market, traders may show the corresponding behaviors. Rockmore and Pauls downloaded the daily close prices of 2547 equities over 1251 days, and searched for some similarities in dynamic prices of different equities. They found 49 clusters of equities, which organized into seven superclusters. These clusters usually corresponded to particular sectors of the economy or to particular places. Superclusters corresponded to industries that depend on one another. These superclusters were connected in a loop which was believed to be the result of a common practice of investment managers (Zimmer, 2011). Rockmore and Pauls used the same methods to build brain model. In this model information flow between a part of the market to other parts is similar to the information flow between a region of the brain to another region. They found 23 clusters in brain, which are further organized in to four superclusters that forms a loop. Their model was expected to push neuroscience towards a truly predictive science (Zimmer, 2011).

Ecological Network Analysis: Innovations

1. Construction of Ecological Networks

Ecological network analysis (ENA) is a methodology to analyze ecological and environmental interactions (Hannon, 1973, 1985, 1986, 1991, 2001; Hannon *et al.*, 1986, 1991; Hannon and Joiris, 1989; Finn, 1976; Patten, 1978, 1981, 1982, 1985; Fath and Patten, 1999a,b; Ulanowicz, 2004; Ulanowicz and Kemp, 1979). ENA quantifies the structure and function of a network by assessing matter and energy flows. It can analyze the structural and functional properties of the network without reducing the model to its presumed minimal constituents (Fath *et al.*, 2007). In a sense the efficiency of transfer, assimilation and dissipation of energy and matter discloses the structure and function of a network.

To conduct ENA, an ecological network should be firstly available. Fath *et al.* (2007) defined a step-by-step procedure for constructing an ecological network and conducting ENA which can be referenced or improved for further application:

(1) **Identify and demarcate the ecosystem of interest.** One should identify the ecosystem of interest and delimit its boundary. Energy-matter transfers within the system constituting the network. Energy-matter transfers across the boundary are either input or output to the network.

(2) **List the major species or functional groups in an ecosystem.** It is necessary to compartmentalize the system into the major groupings. The most aggregated model may have three compartments,

i.e., producers, decomposers/detritus, and consumers. A more dis-aggregated model could have producers, herbivores, carnivores, omnivores, decomposers, and detritivores (Fath, 2004). The most dis-aggregated model is a different compartment for each species. Most models use some aggregation with 6 and 60 compartments.

(3) **Select a unit of currency for the network.** Once the compartments have been chosen, an energy-matter flow currency should be selected. The currency is typically biomass or energy per area for terrestrial and aquatic ecosystems, or volume for aquatic ecosystems per time. In addition to the input, output, and within systems flow transfer values, it is also necessary to measure the mass density (i.e., biomass/area) of each compartment.

(4) **Construct the adjacency matrix to determine any possible flow interactions.** Once the currency has been chosen, one would construct an adjacency matrix to determine whether or not a resource flow of that currency occurs from each compartment to each other one. An adjacency matrix, A, is a representation of the graph structure such that $a_{ij} = 1$ if there is a flow from j to i, else $a_{ij} = 0$. This procedure is to ascertain the possible connectivity of each pair of compartments in the network.

(5) **Empirically measure mass density of every compartment.**

(6) **Empirically measure input, output, and flows between compartments.**

(7) **Use additional knowledge and models to quantify network flows that have not been empirically determined.**

(8) **Employ flow-balancing algorithm to finalize flow matrix and storage, input, and outputs vectors.**

(9) **Apply ENA to network.**

(10) **Sensitivity analysis.**

The data required for ENA should be collected. The contents and procedures are as follows (Fath *et al.*, 2007): The biomass and physiological parameters, such as consumption (C), production (P), respiration (R) and egestion (E) must be quantified for each compartment of the network. Further, the diet of each compartment must be assigned amongst the inputs from other compartments (consumption) of the network. This assignment can be described by a dietary matrix in which the elements denote the matter flows from compartment j to compartment i. For all compartments, inputs should balance outputs ($C = P + R + E$), in accordance with the conservation of matter and the laws of thermodynamics.

Two methods can be used to assign a flow value between compartments:

(1) **MATBLD.** It assigns the flow based on the joint proportion of predator demand and prey availability.
(2) **MATLOD.** It assigns a very small flow to all designated links until either the demand is met or the source is exhausted. The input data for both methods are the biomasses, consumption, production, respiration, egestion, imports and exports of all compartments, and the topology of the networks.

In most cases, a sensitivity analysis is necessary to assess the influences of input data on network results.

Up till now a lot of methods have been used to construct ecological networks. For example, the software package, Ecopath, has more than 2000 registered users in over 120 countries (www.ecopath.org). In inverse modeling (Vezina and Platt, 1988), a food web can be constructed based on available feeding and flow patterns. The biomass of every compartment is assumed to be at a steady state such that the total flows into a compartment are equal to the total amount exiting it. The flows are then back calculated for this particular structure using mass balance equations and basic biological constraints (Fath *et al.*, 2007). The steady-state assumption of this model has some limitations (Vezina and Pahlow, 2003), however, and modifications have been made to relax it (Richardson *et al.*, 2003).

Community assembly rules have been widely used to compensate the lack of empirically derived data in the network construction. In this method, simple algorithms are used to construct hypothetical but ecologically realistic networks. There are two distinct methods with different initial assumptions (Fath *et al.*, 2007):

(1) The first method is based on population/community ecology, which focuses strictly on "who eats whom" and produces structures involving primary producers, grazers, and predators, but explicitly lacks decomposers and detritus. These networks do not typically contain cycling. The two main algorithms of this method are the cascade model

(Cohen and Newman, 1985) and niche model (Williams and Martinez, 2000).

(2) The second method focuses on energy flow in the system and includes all functional groups including detritus and decomposers. The algorithms include a modified niche model (Halnes *et al.*, 2007), a cyberecosystem model (Fath, 2004), and structured food webs of realistic trophic relationships in which transfer coefficients are drawn from uniform and lognormal distributions (Morris *et al.*, 2005).

2. Food Web Models

Some simple models are important to describe non-random food web structure. They include cascade model (Cohen *et al.*, 1990), niche model (Williams and Martinez, 2000), hierarchical model (Cattin *et al.*, 2004), and many others. These static models were based on one-dimensional classification and ordering of species or species hierarchy. Using these models, some properties of food webs can be primarily approached.

2.1. *Multitrophic assembly model*

The multitrophic assembly model was developed by Pimm (1980). Here we describe an extended version of Pimm's multitrophic assembly model (Lockwood *et al.*, 1997). In the extended model, different types of functional responses (i.e., how prey consumption by predators vary with prey density) in predator-prey dynamics were introduced:

(1) Linear functional response (Holling type I), where interaction strength is a per capita consumption rate of prey per predator and thus prey consumption increases linearly with prey abundance, ultimately reaching a maximum for high prey densities.

(2) Prey-dependent functional responses, where interaction strength shows the maximal attack rate of predators on prey, and increases in a decelerating (Holling type II) or sigmoid (Holling type III) fashion, ultimately reaching the maximal attack rate (Holling, 1975). Lotka–Volterra equations were used to model population dynamics (Montoya and Sole, 2003). All model features were predefined in a matrix that represented the regional species pool. Link distributions frequencies in species

pools were constructed randomly (Bollobas, 1985). In each t_i time iterations, one species was randomly chosen and put into the community from the species pool at a population density of 0.001. A species was considered extinct if its density was below that value (Lockwood *et al.*, 1997).

2.2. *Cascade model*

The cascade model (Cohen *et al.*, 1990) is used to theoretically generate a food web. It assigns each species a random value drawn uniformly from the interval [0, 1] and, each species has the probability $P = 2CS/(S-1)$ of consuming only species with the values less than its own. This helps to explain species richness among trophic levels but underestimates interspecific trophic similarity and overestimates length and number of food chains in larger webs.

The following are Java codes, cascadeModel, for the cascade model (Cohen *et al.*, 1990):

```
/*Cascade model of food web generation (Cohen et al., 1990).
Let v be the number of species, d[1 - v][1 - v]: food web matrix,
where if species j feeds on species i, the element (i, j) is 1,
or else 0.*/
  public class cascadeModel {
  public static void main(String[] args){
  int v;
  double c;
  if (args.length!=2)
  System.out.println("You must input the number of species and
directed connectance. For example, you may type the following in
the command window: java cascadeModel 20 0.3, where 20 is the
number of species and 0.3 is directed connectance.");
  v = Integer.valueOf(args[0]).intValue();
  c = Double.valueOf(args[1]).doubleValue();
  cascadeModel(v,c);}
  public static int[][] cascadeModel(int v, double c){
  int i,j,k,r,s;
  double p;
  int d[][] = new int[v + 1][v + 1];
  int b[] = new int[v + 1];
  double x[] = new double[v + 1];
  p = 2*v*c/(v - 1);
  for(i = 1;i< = v;i++)
```

```
    x[i] = Math.random();
    for(i = 1;i <= v;i++) {
    k = 0;
    for(j = 1;j <= v;j++)
    if (x[j] < x[i]) {
    k++;
    b[k] = j; }
    if (Math.random()<=p)
    for(r = 1;r <=k;r++) {
    for(j = 1;j <=v;j++)
    if (b[r] == j) d[i][j] = 1; } }
    System.out.println("Food web generation (Cohen et al., 1990).
There are "+v+" species in the food web.");
    System.out.println("Food web matrix (if species j feeds on
species i, the element (i,j) is 1, or else 0):");
    for(i = 1;i<=v;i++) {
    for(j = 1;j<=v;j++)
    System.out.print(d[j][i]+" ");
    System.out.println(); }
    System.out.println();
    s = 0;
    System.out.println("Number of predators per species
(in-degree):");
    for(i = 1;i <= v;i++) {
    k = 0;
    for(j = 1;j <= v;j++) {
    k+ = d[j][i];
    s+ = d[j][i]; }
    System.out.print(k+" "); }
    System.out.println("\n");
    System.out.println("Number of preys per species
(out-degree):");
    for(i = 1;i <= v;i++) {
    k = 0;
    for(j = 1;j <= v;j++)
    k+ = d[i][j];
    System.out.print(k+" "); }
    System.out.println("\n");
    System.out.println("Total degree of every species:");
    for(i = 1;i <= v;i++) {
    k = 0;
    for(j = 1;j <= v;j++)
    k+ = d[i][j];
    for(j = 1;j <= v;j++)
    k+ = d[j][i];
    System.out.print(k+" "); }
```

```
System.out.println("\n");
System.out.println("There are "+s+" links.");
System.out.println("Directed connectance (C)="+s/Math.pow
(v,2));
return d;}
}
```

2.3. Niche model

2.3.1. Niche model

In the niche model (Williams and Martinez, 2000), each of S species is assigned a "niche value" parameter (n_i), drawn uniformly from the interval $[0, 1]$. Species i consumes all species falling in a range (r_i) in which the center (c_i) is uniformly drawn from $[r_i/2, n_i]$. It helps to produce loops and cannibalism by increasing up to half of r_i to include values $> n_i$. The value of r_i is determined by using a beta function to randomly draw values from $[0, 1]$ whose expected value is $2C$ and then multiplying the value by n_i (expected $E(n_i) = 0.5$) to obtain desired C. A beta distribution with $\alpha = 1$ is used. It has the form $f(x|1, \beta) = \beta(1-x)^{\beta-1}$, and the expectation $E(x) = 1/(1+\beta)$. In this case, $x = 1 - (1-y)^{1/\beta}$ is a random variable from the beta distribution if y is a uniform random variable and b is chosen to obtain the desired expected value. The fundamental generality of species i is measured by r_i. The number of species falling within r_i represents realized generality. Sometimes food webs generated by models contain completely disconnected species (isolated species) or trophically identical species. All such species should be eliminated and replaced. The species with the smallest n_i has $r_i = 0$ so that every food web has at least one basal species.

2.3.2. Variant model

The preceding niche model assigns each species a randomly drawn "niche value". The species are then constrained to consume all prey species within one range of values whose randomly chosen center is less than the consumer's niche value. A variant of niche model was thus presented (Williams and Martinez, 2000).

According to the variant of niche model, there are two steps in the construction of food web. First, randomly assign species $i = 1, 2, \ldots, s$, to the niche nutrients in the interval $[0, 1]$. Second, randomly assign species *i* into the interval $[0, 1]$; all species within the interval are considered to be preys of species *i*. The food web matrix is thus constructed.

Trophic similarity of a pair of species (s_{ij}) is the number of predators and prey shared in common divided by the pair's total number of predators and prey. Mean maximum similarity of a web can be calculated by averaging all species' largest similarity index (Williams and Martinez, 2000).

The following are Java codes, nicheModel, for the variant of niche model of Williams and Martinez (2000):

```
/*Niche model of food web generation (Williams and Martinez,
2000). Let v be the number of species, d[1 - v][1 - v] is food web
matrix, where if species j feeds on species i, the element (i,j)
is 1, or else 0.*/
    public class nicheModel {
    public static void main(String[] args){
    int v;
    if (args.length!=1)
    System.out.println("You must input the number of species. For
example, you may type the following in the command window: java
nicheModel 20, where 20 is the number of species.");
    v = Integer.valueOf(args[0]).intValue();
    nicheModel(v);
    nicheModelVariant(v); }
    public static int[][] nicheModelVariant(int v) {
    int i,j,k,s;
    double a,b,c;
    int d[][] = new int[v + 1][v + 1];
    double x[] = new double[v + 1];
    for(i = 1;i <= v;i++)
    x[i] = Math.random();
    for(i = 1;i <= v;i++) {
    a = x[i]*Math.random();
    b = x[i]*Math.random();
    if (b < a) {
    c = a;
    a = b;
    b = c; }
    for(j = 1;j <= v;j++) {
    if (i==j) continue;
    if ((x[j]>=a) & (x[j]<=b))
    d[i][j] = 1; } }
    System.out.println("Variant model of food web generation
(Williams and Martinez,2000).");
    print(v, d);
```

```
speciestype(v, d);
tropsimil(v, d);
return d; }
public static int[][] nicheModel(int v) {
int i,j;
double a,b,c,r;
int d[][] = new int[v + 1][v + 1];
double x[] = new double[v + 1];
for(i = 1;i<=v;i++)
x[i] = Math.random();
for(i = 1;i <= v;i++) {
r = range(x[i]);
c = Math.random();
c = x[i]*c + r/2*(1 - c);
a = c - r/2;
b = c + r/2;
for(j = 1;j<=v;j++) {
if (i==j) continue;
if ((x[j]>=a) & (x[j]<=b))
d[i][j] = 1; } }
System.out.println("Food web generation (Williams and
Martinez,2000).");
print(v,d);
speciestype(v,d);
tropsimil(v,d);
return d; }
public static double range(double x) {
int i,max=1000;
double s,y,xx,r,mean,beta;
s = 0;
for(i = 1;i<=max;i++) {
y=Math.random();
s+ = y; }
mean = s/max;
beta = 1/mean-1;
s = 0;
for(i = 1;i<=max;i++) {
xx = 1 - Math.pow(1 - Math.random(),1/beta);
s+ = xx; }
mean = s/max;
r = mean*x;
return r; }
public static void speciestype(int v, int d[][]) {
int i,j,s;
int a[] = new int[v+1];
int b[] = new int[v+1];
```

```
for(i = 1;i<=v;i++) {
a[i] = b[i] = 0;
for(j = 1;j<=v;j++) {
a[i]+ = d[j][i];
b[i]+ = d[i][j]; } }
s = 0;
System.out.print("Top species:");
for(i = 1;i<=v;i++)
if (a[i]==0) {
System.out.print(i+" ");
s++;}
System.out.println();
System.out.println("Total number:"+s);
System.out.println();
s = 0;
System.out.print("Intermediate species:");
for(i = 1;i<=v;i++)
if ((a[i]!=0) & (b[i]!=0)) {
System.out.print(i+" ");
s++;}
System.out.println();
System.out.println("Total number:"+s);
System.out.println();
s = 0;
System.out.print("Basal species:");
for(i = 1;i<=v;i++)
if (b[i]==0) {
System.out.print(i+" ");
s++;}
System.out.println();
System.out.println("Total number:"+s);
System.out.println(); }
public static double[][] tropsimil(int v, int d[][]) {
int i,j,k,c,cc;
double max,sum,mxsim;
double s[][] = new double[v+1][v+1];
mxsim = sum = 0;
for(i = 1;i<=v;i++) {
max = 0;
for(j = 1;j<=v;j++) {
if (i==j) {
s[i][j] = 1;
continue; }
c = cc = 0;
for(k = 1;k<=v;k++) {
if ((d[i][k]==1) & (d[j][k]==1)) c++;
```

```
if ((d[i][k]==1) | (d[j][k]==1)) cc++; }
for(k = 1;k<=v;k++) {
if ((d[k][i]==1) & (d[k][j]==1)) c++;
if ((d[k][i]==1) | (d[k][j]==1)) cc++; }
s[i][j]=c*1.0/cc;
if (s[i][j]>max) max=s[i][j]; }
sum+ = max; }
mxsim = sum/v;
System.out.println("Trophic similarity of paired
species:");
for(i = 1;i<=v;i++) {
for(j = 1;j<=v;j++)
System.out.print(s[i][j]+" ");
System.out.println(); }
System.out.println();
System.out.println("Maximum similarity of the web:"+mxsim);
System.out.println();
return s; }
public static void print(int v, int d[][]) {
int i,j,k,s;
double c;
System.out.println("There are "+v+" species
in the food web.");
System.out.println("Food web matrix (if species j feeds on
species i, the element (i,j) is 1, or else 0):");
for(i = 1;i<=v;i++) {
for(j = 1;j<=v;j++)
System.out.print(d[j][i]+" ");
System.out.println(); }
System.out.println();
s = 0;
System.out.println("Number of predators per species
(in-degree):");
for(i = 1;i<=v;i++) {
k = 0;
for(j = 1;j<=v;j++) {
k+=d[j][i];
s+=d[j][i]; }
System.out.print(k+" "); }
System.out.println("\n");
System.out.println("Number of preys per species
(out-degree):");
for(i = 1;i<=v;i++) {
k = 0;
for(j = 1;j<=v;j++)
k+ = d[i][j];
```

```
System.out.print(k+" "); }
System.out.println("\n");
System.out.println("Total degree of every species:");
for(i = 1;i<=v;i++) {
k = 0;
for(j = 1;j<=v;j++)
k+ = d[i][j];
for(j = 1;j<=v;j++)
k+=d[j][i];
System.out.print(k+" "); }
System.out.println("\n");
System.out.println("There are "+s+" links.");
c = s/Math.pow(v,2);
System.out.println("Directed connectance (C)="+c);
System.out.println("\n"); }
}
```

Niche model performs excellently in predicting degree distribution of food webs, in particular the distribution of total degree and outdegree. Moreover, it will help to develop a general function for degree distribution.

In the niche model, species are randomly assigned to an interval. This is reasonable in some cases. However, the assignment is not accordant with true situations if between-species trophic similarity is high. In these situations species are not randomly but intensively distributed over the interval $[0, 1]$, which results in the increase of species with high links compared with the prediction of niche model. Diverse species aggregate with similar trophic requirements and are thus different from the ones predicted by niche model.

2.4. *MaxEnt model*

A model, MaxEnt, which is based on the maximum-entropy principle, can be used to predict degree distribution of ecological network (Williams, 2010).

MaxEnt model includes prior knowledge of the number of basal species B and does not attempt to predict the fraction of basal species. Similarly, two consumer distributions are considered, the "all-species consumer distribution" and the "restricted consumer distribution." The "all-species consumer distribution" is defined as the distribution of the number of consumers of each species, including the top species, which have no consumers. The "restricted consumer distribution" is defined as the distribution of the number of consumers of the resource species, which includes prior knowledge of the number of top species T, and does not attempt to predict the fraction of top species (Williams, 2010).

In the all-species consumer distribution, the resource species or consumers of each species can range from zero to S (S: number of species; L: links), and the mean number of links per species is L/S. In the restricted resource distribution, the number of links per consumer varies within one and S, and the mean number of links per consumer is $L/(S - B)$. In the restricted consumer distribution, the number of links from each resource varies within one and S, and the mean number of links from each resource is $L/(S - T)$.

First, a discrete distribution should be found based on some values, for example $\{x_1, \ldots, x_n\}$. The mean μ (i.e., indegree or outdegree of a node) maximizes $H = -\sum_i p_i \ln p_i$, and satisfies some conditions. The maximum entropy distribution is

$$p_i = P(X = x_i) = Ce^{\lambda x_i}, \quad \text{for } i = 1, \ldots, n.$$

Williams (2010) used the model $p_i = P(X = x_i) = x_i^\lambda$, i.e., using power law distribution but not exponential distribution. Set the conditions (Jaynes, 1957), $\sum_i P_i = 1$ and $\sum_i X_i P_i = \mu$. The values for C and λ can thus be obtained by using Lagrange multipliers. Williams (2010) assumes that the number of consumers and resources are independent on each node of food web. For T top species without consumers, the number of links is from maximum entropy resource distribution. Similarly, for B basal species, the number of links is from maximum entropy consumer distribution and for $S - B - T$ intermediate species, the number of links is the sum of the numbers from consumer and resource distributions.

In addition, Simpson index can also be used as the entropy index:

$$H = 1 - \sum p_i^2$$

As a supplement, some other degree distributions are also described as follows.

Barabasi (2009):

$$f(r) = r^{-\beta}$$

Zhang (2011a):

$$f(r) = \alpha(r-a)^{\beta}(b-r)^{\eta}, \quad a \le r \le b;$$
$$f(r) = 0, \quad r > a \ \text{ or } \ r < b;$$
$$\alpha > 0, \quad \beta \ge 0, \quad \eta \ge 0.$$

where α: scale parameter; β, γ: shape parameters; a, b: position parameters.

The following are Java codes, maxEnt, for maximum entropy algorithm:

```
/*Maximum entropy algorithm for network generation. In the
basal species connection number file (maxent1) and top species
connection number file (maxent2), the numbers of basal and top
species connections of the network are given respectively.*/
  public class maxEnt {
  static double x[] = new double[10];
  public static void main(String[] args){
  int i,j,vth,vt,vb,nf1,nf2,typedis,typeent;
  double miuin,miuout;
  if (args.length!=11) System.out.println("You must input the
names of basal and top species connection number tables in the
database, the different number of connections in basal species
connection number table, the different number of connections in
top species connection number table, the expected mean number of
basal species connections, the expected mean number of top
species connections, the type of frequency distribution of
degrees (1: exponential distri (p(x)=ce^(dx)); 2: power law
distri (p(x)=x^(-c)); 3: Zhang's distri (p(x)=a(x - b)^c*
(d - x)^e)), the type of entropy expression (1: Shannon - Wiener
index; 2: Simpson index), the number of total species, the number
of top species, and the number of basal species. For example, you
may type the following in the command window: java maxEnt maxent1
maxent2 4 7 5 8 1 2 15 3 6, where maxent1 and maxent2 are the names
of basal and top species connection number tables respectively,
there are 4 and 7 different numbers of basal and top species
connections in the two tables respectively, there are totally
15 species in which there are three top species and six basal
species, 1 means exponential distribution, 2 means Simpson index,
the expected mean numbers of basal and top species connections are
5 and 8 respectively.");
  String tablename1 = args[0];
  String tablename2 = args[1];
  nf1 = Integer.valueOf(args[2]).intValue();
  nf2 = Integer.valueOf(args[3]).intValue();
```

```
miuin = Double.valueOf(args[4]).doubleValue();
miuout = Double.valueOf(args[5]).doubleValue();
typedis = Integer.valueOf(args[6]).intValue();
typeent = Integer.valueOf(args[7]).intValue();
vtb = Integer.valueOf(args[8]).intValue();
vt = Integer.valueOf(args[9]).intValue();
vb = Integer.valueOf(args[10]).intValue();
readDatabase                              readdata1=new
readDatabase("dataBase",tablename1,nf1);
readDatabase                              readdata2=new
readDatabase("dataBase",tablename2,nf2);
double ff1[] = new double[nf1+1];
double ff2[] = new double[nf2+1];
for(j = 1;j<=nf1;j++)
ff1[j]=(Integer.valueOf(readdata1.data[1][j])).intValue();
for(j = 1;j<=nf2;j++)
ff2[j]=(Integer.valueOf(readdata2.data[1][j])).intValue();
maxent(typedis,typeent,nf1,nf2,miuin,miuout,vtb,vt,
vb,ff1,ff2);}
public static void maxent(int typedis, int typeent, int nf1,
int nf2, double miuin, double miuout, int v, int vt, int vb,
double ff1[], double ff2[])
{
  int i,j,nf,k = 0;
  double ran,mis,mos;
  double xi[] = new double[nf1+nf2+1];
  double p[] = new double[nf1+nf2+1];
  double u[] = new double[nf1+nf2+2];
  int t[] = new int[vt+1];
  int b[] = new int[vb+1];
  nf = nf1;
  for(i = 1;i<=nf;i++)
  xi[i] = ff1[i];
  System.out.println("Degree  distribution  of  basal
species:");
  simplex(typedis,typeent,nf,miuin,xi);
  for(i = 1;i<=nf;i++)
  if (typedis==1) p[i]=x[1]*Math.exp(x[2]*xi[i]);
  else if (typedis==2) p[i]=Math.pow(xi[i],-x[1]);
  else if (typedis==3)
p[i]=x[1]*Math.pow(xi[i]-x[2],x[3])*Math.pow(x[4]-xi[i],
x[5]);
  u[0] = 0;
  for(i = 1;i<=nf;i++)
  u[i]=u[i - 1]+p[i];
  for(i = 1;i<=vb;i++) {
```

```
ran = Math.random()*u[nf];
for(j = 1;j<=nf;j++)
if ((ran>=u[j - 1]) & (ran<u[j])) {
k = j;
break;}
b[i] = (int)xi[k];}
nf = nf2;
for(i = 1;i<=nf;i++)
xi[i] = ff2[i];
System.out.println("Degree distribution of top species:");
simplex(typedis,typeent,nf,miuout,xi);
for(i=1;i<=nf;i++)
if (typedis==1) p[i] = x[1]*Math.exp(x[2]*xi[i]);
else if (typedis==2) p[i] = Math.pow(xi[i],-x[1]);
else if (typedis==3) p[i]=x[1]*Math.pow(xi[i] -
x[2],x[3])*Math.pow(x[4] - xi[i],x[5]);
u[0] = 0;
for(i = 1;i<=nf;i++)
u[i]=u[i -1]+p[i];
for(i = 1;i<=vt;i++) {
ran = Math.random()*u[nf];
for(j = 2;j<=nf;j++)
if ((ran>=u[j - 1]) & (ran<u[j])) {
k = j;
break; }
t[i] = (int)xi[k]; }
mis = mos = 0;
for(i = 1;i<=vt;i++)
mis+ = t[i];
for(i = 1;i<=vb;i++)
mos+ = b[i];
System.out.println("Food web generation (Williams, 2010;
Barabasi, 2009; Zhang, 2011a)");
System.out.println("Indegree of "+vb+" basal species:");
for(i = 1;i<=vb;i++)
System.out.print(b[i]+" ");
System.out.println();
System.out.println("Total degree of "+(v - vt - vb)+"
intermediate species: "+(mos+mis));
System.out.println("Outdegree of "+vt+" top species:");
for(i = 1;i<=vt;i++)
System.out.print(t[i]+" ");
System.out.println(); }
public static double[] simplex(int typedis, int typeent,
int nf, double miu, double xi[]) {
double dt=0,fm=0,pr,t,gj,al,be,ga,sq,tk,d1,d2,fh,fh2,fl,
```

```
ub,ua,ui,uj;
  int i,j,n = 2,max,it,n1,n2,n3,n4,n5,h1,l1;
  max = 50000;
  pr = 1e - 05;
  if (typedis==1) n = 2;
  if (typedis==2) n = 1;
  if (typedis==3) n = 5;
  double xl[][] = new double[n+7][n+2];
  double d[][] = new double[n+3][n+2];
  double f[] = new double[n+7];
  t = 0.01;
  al = 0.01;
  be = 0.01;
  ga = 0.01;
  n1 = n2 = n3 = n4 = n5 = n+1;
  sq = Math.sqrt(n1);
  tk = t/(n*Math.sqrt(2));
  d1 = tk*(sq+n-1);
  d2 = tk*(sq-1);
  for(i = 1;i<=n;i++) {
  if (typedis==1) {
  x[1] = Math.random();
  x[2] = - 1+Math.random(); }
  else if (typedis==2) x[1] = 1+Math.random();
  if (typedis==3) {
  x[1] = Math.random();
  x[2] = Math.random();
  x[3] = 2+Math.random();
  x[4] = xi[nf]+Math.random();
  x[5] = 2+Math.random(); } }
  gj = objectivefun(typedis,typeent,nf,miu,xi);
  for(j = 1;j<=n;j++) d[1][j]=0;
  for(i = 2;i<=n1;i++) {
  for(j = 1;j<=n;j++) d[i][j]=d2;
  d[i][i - 1] = d1; }
  for(i = 1;i<=n1;i++)
  for(j = 1;j<=n;j++) xl[i][j] = x[j]+d[i][j];
  for(i = 2;i<=n1;i++) {
  for(j = 1;j<=n;j++) x[j] = xl[i][j];
  f[i] = objectivefun(typedis,typeent,nf,miu,xi); }
  it = 0;
  point1: while (it>=0) {
  fh = fl = gj;
  h1 = l1 = 1;
  for(i = 2;i<=n1;i++)
  if (f[i]<=fh) {
```

```
if (f[i]>=fl) continue;
else fl = f[i];
l1 = i;
continue; }
else {
fh = f[i];
h1 = i;
if (f[i]>=fl) continue;
else fl = f[i];
l1 = i;
continue; }
for(j = 1;j<=n;j++) {
xl[n2][j] = 0;
for(i = 1;i<=n1;i++) xl[n2][j]+=xl[i][j];
xl[n2][j] = (xl[n2][j] - xl[h1][j])/n;
xl[n3][j] = (1+al)*xl[n2][j] - al*xl[h1][j];
x[j] = xl[n3][j]; }
ub = objectivefun(typedis,typeent,nf,miu,xi);
if (ub<fl) {
for(j=1;j<=n;j++) {
xl[n4][j] = (1 - al)*xl[n2][j]+ga*xl[n3][j];
x[j] = xl[n4][j]; }
ua = objectivefun(typedis,typeent,nf,miu,xi);
if (ua>=fl) {
for(j=1;j<=n;j++) xl[h1][j]=xl[n3][j];
f[h1] = ub;
for(j = 1;j<=n;j++) x[j]=xl[n2][j];
ui = objectivefun(typedis,typeent,nf,miu,xi);
dt = 0;
for(i = 1;i<=n1;i++) dt+=Math.pow(f[i]-f[l1],2);
dt = Math.sqrt(dt);
if ((it>=max) | (dt<pr)) {
for(j=1;j<=n;j++) x[j]=xl[l1][j];
fm = f[l1];
break point1; }
else {
it++;
continue point1; } }
else {
for(j = 1;j<=n;j++) xl[h1][j]=xl[n4][j];
f[h1] = ua;
for(j = 1;j<=n;j++) x[j]=xl[n2][j];
ui = objectivefun(typedis,typeent,nf,miu,xi);
dt = 0;
for(i = 1;i<=n1;i++) dt+=Math.pow(f[i] - f[l1],2);
dt = Math.sqrt(dt);
```

```
if ((it>=max) | (dt<pr)) {
for(j = 1;j<=n;j++) x[j] = xl[l1][j];
fm = f[l1];
break point1; }
else {
it++;
continue point1; } } }
fh2 = gj;
if (h1==1) fh2=f[2];
for(i = 2;i<=n2;i++) {
if ((i==h1) | (f[i]<=fh2)) continue;
else fh2 = f[i]; }
if (ub<=fh2) {
for(j = 1;j<=n;j++) xl[h1][j] = xl[n3][j];
f[h1] = ub;
for(j = 1;j<=n;j++) x[j] = xl[n2][j];
ui = objectivefun(typedis,typeent,nf,miu,xi);
dt = 0;
for(i = 1;i<=n1;i++) dt+=Math.pow(f[i]-f[l1],2);
dt = Math.sqrt(dt);
if ((it>=max) | (dt<pr)) {
for(j = 1;j<=n;j++) x[j]=xl[l1][j];
fm = f[l1];
break point1; }
else {
it++;
continue point1; } }
if (ub<=fh)
for(j = 1;j<=n;j++) xl[h1][j] = xl[n3][j];
for(j = 1;j<=n;j++) {
xl[n5][j] = be*xl[h1][j]+(1 - be)*xl[n2][j];
x[j] = xl[n5][j]; }
uj = objectivefun(typedis,typeent,nf,miu,xi);
if (uj<=fh) {
for(j = 1;j<=n;j++) xl[h1][j] = xl[n5][j];
f[h1] = uj;
for(j = 1;j<=n;j++) x[j] = xl[n2][j];
ui = objectivefun(typedis,typeent,nf,miu,xi);
dt = 0;
for(i = 1;i<=n1;i++) dt+=Math.pow(f[i]-f[l1],2);
dt = Math.sqrt(dt);
if ((it>=max) | (dt<pr)) {
for(j = 1;j<=n;j++) x[j]=xl[l1][j];
fm = f[l1];
break point1; }
else {
```

```
it++;
continue point1; } }
for(i = 1;i<=n1;i++) {
for(j = 1;j<=n;j++) {
xl[i][j] = t*(xl[i][j]+xl[l1][j]);
x[j] = xl[i][j]; }
f[i] = objectivefun(typedis,typeent,nf,miu,xi); }
for(j = 1;j<=n;j++) x[j]=xl[n2][j];
ui = objectivefun(typedis,typeent,nf,miu,xi);
dt = 0;
for(i = 1;i<=n1;i++) dt+=Math.pow(f[i] - f[l1],2);
dt = Math.sqrt(dt);
if ((it>=max) | (dt<pr)) {
for(j = 1;j<=n;j++) x[j] = xl[l1][j];
fm = f[l1];
break point1; }
else {
it++;
continue point1; } }
if (typedis==1)
System.out.println("p(x)="+x[1]+"exp("+x[2]+"x)");
   if (typedis==2) System.out.println("p(x)="+"x^
"+(-x[1]));
   if (typedis==3)
System.out.println("p(x)="+x[1]+"(x-"+x[2]+")^
"+x[3]+"("+x[4]+"-x)^"+x[5]);
return x; }
public static double obj(int term, int nf, double p[],
double xi[]) {
double s=0;
for(int j=1;j<=nf;j++)
switch (term) {
   case 1: s+ = p[j]; break;
   case 2: s+ = p[j]*xi[j]; break;
   case 3: s+ = Math.log(p[j])*p[j]; break;
   case 4: s+ = Math.pow(p[j],2); break; }
return s; }
public static double objectivefun(int typedis,
int typeent, int nf, double miu, double xi[]) {
double f = 0;
double p[] = new double[nf+1];
for(int i = 1;i<=nf;i++)
if (typedis==1) p[i]=x[1]*Math.exp(x[2]*xi[i]);
else if (typedis==2) p[i]=Math.pow(xi[i], - x[1]);
else if (typedis==3) p[i]=x[1]*Math.pow(xi[i]-x[2],x[3])*
```

```
Math.pow(x[4] - xi[i],x[5]);
  switch (typeent) {
  case 1:
f=obj(3,nf,p,xi)+Math.abs(obj(1,nf,p,xi)-1)+
Math.abs(obj(2,nf,p,xi)-miu);
  break;
  case 2:
f=obj(4,nf,p,xi)-1+Math.abs(obj(1,nf,p,xi)-1)+
Math.abs(obj(2,nf,p,xi)-miu);
  break; }
  return f; }
  }
```

Resource distributions of consumers and various trophic levels for 51 food webs have been analyzed and clarified (Cohen *et al.*, 1990).

Degree distributions of the 51 empirical food webs and the corresponding prediction of maximum entropy model are consistent. This demonstrates that it is not necessary to consider the detailed ecological process in many food webs in order to predict the (undirected) degree distributions of consumers and resources.

2.5. *Ecopath model*

Ecopath is a free software to construct and analyze mass-balanced flow networks. It is a software based on a method proposed by Polovina (1984), and upgraded with some more methods (Christensen and Pauly, 1992; Walters *et al.*, 1997, 2000; Pauly *et al.*, 2000; Christensen and Walters, 2004). Ecopath is the core program of the EwE (Ecopath with Ecosim) (Libralato *et al.*, 2006). The interactions in the networks represent trophic links at the species or functional level. It allows users to input ecosystem data such as the total mortality estimates, consumption estimates, diet compositions, and fishery catchers (Christensen and Walters, 2004). The model uses two main equations, one regarding production (production = catch + predation + net migration + biomass accumulation + other mortality), and the other regarding consumption (consumption = production + respiration + unassimilated food).

According to Libralato *et al.* (2006), the balance of mass (energy or nutrients) of any functional group *i* of the network can be achieved by setting its production equal to the sum of the consumption terms:

$$(P/B)_i B_i = \sum_{j=1}^{n} (Q/B)_j B_j DC_{ij} + E_i + Y_i + BA_i + (P/B)_i B_i (1 - EE_i)$$

where production (i.e., the left side of the equation) is the product of the production *vs.* biomass ratio, $(P/B)_i$, and the biomass, B_i; the right-hand side terms are the sum of the predation terms: the product of the consumption *vs.* biomass ratio, $(Q/B)_j$, the biomass of the predators, B_j, and the proportion of the prey i in the diet of the predator j, DC_{ij}; the net flow through the boundaries of the system, E_i; the fishing exploitation, Y_i; the accumulation or depletion of biomass, BA_i; and the last term, non-predation natural mortality, where EE_i is the ecotrophic efficiency.

The flows within a trophic web can be achieved by solving the above equations.

Given the mass balance model of a trophic network, the mixed trophic impact of each pair of functional groups (i, j) in the network, m_{ij} (direct and indirect impacts that each group i has on any group j of the network. It is the first order partial derivatives (in terms of biomass of the master equation above) can be estimated based on the net impact matrix (Libralato *et al.*, 2006). Positive (negative) values of m_{ij} represent the increase (decrease) of biomass of the group j due to a slight increase of biomass of the group i.

First, the net impact of i on j, q_{ij}, is given by the difference between positive effects, quantified by the fraction of the prey i in the diet of the predator j, d_{ji}, and negative effects, evaluated through the fraction of total consumption of i used by predator j, f_{ij} (Ulanowicz and Puccia, 1990). By doing so, the resulting matrix of the net impacts, $Q = (q_{ij})$ is thus:

$$q_{ij} = d_{ji} - f_{ij}$$

The mixed trophic impact m_{ij} is the product of all the net impacts q_{ij} for all the possible pathways in the network that link the functional groups i and j. As an easier way, the matrix of the mixed trophic impacts, M, can be obtained by the inverse of the matrix Q (Ulanowicz and Puccia, 1990).

According to Libralato *et al.* (2006), negative elements of matrix M indicate a prevailing of negative effects, or the effects of the predator on the prey; and positive elements of M indicate prevailing effects of the prey on the predator. Negative elements of M can thus be associated to prevailing top-down effects and positive ones to bottom-up effects.

2.6. *Ecosim method*

Ecosim is the dynamic program of the EwE (Walters *et al.*, 1997, 2000). It is based on a group of differential equations derived from the Ecopath equation above. This allows a dynamic representation of the system variables like biomasses, predation, and production, etc (Libralato *et al.*, 2006). Ecosim uses Lotka–Volterra models that have been modified to include foraging arena theory (Walters *et al.*, 1997, 2000), which avoids Lotka–Volterra's unrealistic assumption of uniform or random distribution of species interactions. In foraging arena theory, the biomass of the prey available to predators is only a vulnerable fraction of total biomass, with exchanges rates between the vulnerable and the invulnerable states calculated using vulnerability coefficients (Christensen and Walters, 2004; Libralato *et al.*, 2006).

The Ecosim master equation takes the following form (Pinnegar *et al.*, 2005):

$$dB_i/dt = g_i \sum_{j=1} Q_{ji} - \sum_{j=1} Q_{ij} + I_i - (M_i + F_i + e_i)B_i$$

where dB_i/dt is the growth rate of group i during the time interval dt, g_i is the net growth efficiency (production/consumption ratio), M_i is the non-predation natural mortality rate, F_i is the fishing mortality rate, e_i is the emigration rate, I_i is the immigration rate. In this equation, consumption rates, Q_{ji}, are calculated based on a "foraging arena" idea, in which the B_i's are divided into vulnerable and invulnerable components (Walters *et al.*, 1997). The size of the transfer rate, v_{ij}, i.e., the rate at which prey move between being "vulnerable" and "not vulnerable", determines if control is top-down, bottom-up or an intermediate type.

3. Indices for Keystone Species in Food Web

3.1. *Centrality indices*

Centrality indices can be used to describe ecological network (Navia *et al.*, 2010).

The first centrality index is betweenness centrality. It measures how central a given vertex is in terms of being adjacent to many shortest paths in the network. It is based on quantifying how often vertex i is on the shortest path between each pair of vertices j and k. The standardized centrality index for vertex i is:

$$C_i = 2 \sum_{j \leq k} g_{jk}(i)/g_{jk}/[(N-1)(N-2)]$$

where $i \neq j$ and k, g_{jk} is the number of equally shortest paths between vertices j and k, and $g_{jk}(i)$ is the number of these shortest paths to which vertex i is adjacent. The denominator is twice the number of pairs of vertices without vertex i. If C_i is large for trophic group i, the loss of this vertex will have many rapidly spreading effects in the web.

The second centrality index is closeness centrality. It measures the closeness of a vertex to rest of the vertices. It is based on the proximity principle and quantifies how short the minimal paths from a given vertex to all other vertices are (Wassermann and Faust, 1994). The standardized form is:

$$CC_i = (N-1)/\sum_{j=1}^{N} d_{ij}$$

where $i \neq j$, and d_{ij} is the length of the shortest path between vertices i and j in the network. The smallest value of CC_i will be for that of the trophic group which upon being removed will affect the majority of other groups.

3.2. *Keystone index*

In 1969 Paine proposed the "keystone species" concept. He defined keystone species as such predators: They can control the population density of prey below the resource limitation level and prevent prey population from extinction. It stresses the mechanism of top-down control. Mills argued that the keystone species have two distinct characteristics: (1) Their existence is so important in preserving composition, structure and diversity of biological community. (2) These species are closely related to other members in the community.

Menge defined keystone species as the predator that independently determines the structural pattern (distribution, abundance, composition, diversity, etc.) of preys.

Keystone index is usually used to characterize the importance of species in ecosystems according to their position in the trophic network. It is also known as the index of topological importance.

The concept of keystone index proposed by Jordon (2001) and Jordon *et al.* (2006) was based on food web, which can be expressed as:

$$K_j = \sum_{c=1}^{n}(1 + K_{bc})/d_c + \sum_{e=1}^{m}(1 + K_{te})/f_e$$

where n is the number of predators eating species i, d_c is the number of prey of the cth predator and K_{bc} is the bottom-up keystone index of the cth predator. Similarly, m is the number of prey eaten by species i, f_e is the number of predators of the eth prey and K_{te} is the top-down keystone index of the eth prey.

This index is characterized by the following features: (1) Including the between-species effects at the same trophic level. (2) Considering both top-down and bottom-up relations of species. (3) Species only within the same food web can be compared for their relative importance. The comparison of species between different food webs is not allowed.

4. Flow Indices and Matrices of Network

Some standardized indices and matrices (Latham, 2006) for ecological networks are described as below. Latham (2006) has provided a free BASIC software package, NetMatCalc, for easy calculation of these indices or matrices. For more details, examine the review of Latham (2006).

4.1. *Average mutual information*

Average Mutual Information (AMI) (Rutledge *et al.*, 1976) is used to measure the average amount of constraint placed on a unit of flow anywhere in the network (Latham and Scully, 2002; Latham, 2006).

AMI assumed that rigid diets and highly constrained flow in an ecosystem would result in a fragile system unable to persist in the changing environmental conditions (Rutledge *et al.*, 1976). As a system grows and matures to form a network of energy and material flow, the AMI should be

(Latham, 2006):

$$\text{AMI} = k \sum_{i=1}^{n+2} \sum_{j=0}^{n} (T_{ij}/T_{..}) \log_2 (T_{ij} T_{..}/(T_{i.}.T_{.j})),$$

where n: the number of nodes or compartments in the network (but does not include compartment 0, $n + 1$, or $n + 2$); k: constant scalar; T_{ij}: flow from compartments j to i; $T_{i.}$: total inflow for compartment i; $T_{.j}$: total outflow for compartment j; $T_{..}$: total system throughput, i.e., the sum of network links:

$$T_{..} = \sum_{i=1}^{n+2} \sum_{j=0}^{n} T_{ij}$$

The upper bound of AMI is:

$$H_R = - \sum_{j=0}^{n} (T_{.j}/T_{..}) \log_2 (T_{.j}/T_{..})$$

AMI/H_R is suggested to be useful for comparing the degree of constraint across systems (Latham, 2006).

4.2.   *Ascendency*

Ascendency index (A) was developed by Ulanowicz *et al.* (1990, 1997):

$$A = \text{AMI} * T_{..}$$

AMI was believed to increase during a system's development toward a climax community, due to the pruning of less efficient pathways in the network and the increasing autocatalytic between-compartment interconnection (Ulanowicz, 1980, 1997).

A suggested measure for comparing ascendency across networks is the development ascendency, which is represented by (Latham 2006):

$$T_{..} = -A \Big/ \left[\sum_{i=1}^{n+2} \sum_{j=0}^{n} T_{ij} \log_2 (T_{ij}/T_{..}) \right]$$

Ulanowicz and Norden (1990) provided four types of measures to study specific aspects of networks (Latham, 2006):

Network import:

$$A_0 = \sum_{i=1}^{n} T_{i0} \log_2(T_{i0} T_{..}/(T_{i.} T_{.0}))$$

$$\Phi_0 = -\sum_{i=1}^{n} T_{i0} \log_2(T_{i0}^2/(T_{i.} T_{.0}))$$

$$C_0 = -\sum_{i=1}^{n} T_{i0} \log_2(T_{i0}/T_{..})$$

Network internal:

$$A_I = \sum_{i,j=1}^{n} T_{ij} \log_2(T_{ij} T_{..}/(T_{i.} T_{.j}))$$

$$\Phi_I = -\sum_{i,j=1}^{n} T_{ij} \log_2(T_{ij}^2/(T_{i.} T_{.j}))$$

$$C_I = -\sum_{i,j=1}^{n} T_{ij} \log_2(T_{ij}/T_{..})$$

Network export:

$$A_E = \sum_{j=1}^{n} T_{n+1j} \log_2(T_{n+1j} T_{..}/(T_{n+1.} T_{.j}))$$

$$\Phi_E = -\sum_{j=1}^{n} T_{n+1j} \log_2(T_{n+1j}^2/(T_{n+1.} T_{.j}))$$

$$C_E = -\sum_{j=1}^{n} T_{n+1j} \log_2(T_{n+1j}/T_{..})$$

Network dissipation:

$$A_S = \sum_{j=1}^{n} T_{n+2j} \log_2(T_{n+2j} T_{..}/(T_{n+2.} T_{.j}))$$

$$\Phi_S = -\sum_{j=1}^{n} T_{n+2j} \log_2(T_{n+2j}^2/(T_{n+2}.T._j))$$

$$C_S = -\sum_{j=1}^{n} T_{n+2j} \log_2(T_{n+2j}/T..)$$

4.3. *Compartmentalization index*

Compartmentalization index (C) is used to measure the degree of well connected subsystems (which should have similar number of species and connectance) in a network (Pimm and Lawton, 1980):

$$C = \left[\sum_{i=1}^{n} \sum_{j\neq ij=1}^{n} c_{ij}/(n(n-1)) \right]$$

where c_{ij}: the number of species with which both i and j interact divided by the number of species with which either i or j interact.

A higher value of C means a higher compartmentalization of the network.

4.4. *Constraint efficiency*

Constraint efficiency (CE) is a measure of total of constraints that govern flow out of individual compartments (Latham and Scully, 2002):

$$CE = \left[\sum_{i=1}^{n} \log_2(n+2) + \sum_{i=1}^{n+2} \sum_{j=1}^{n} (T_{ij}/T._j) \log_2(T_{ij}/T._j) \right] \Bigg/ \sum_{i=1}^{n} \log_2(n+2)$$

Constraint efficiency is the sum of all constraints. Thus local changes of network topography will be directly reflected in the system-wide value of constraint efficiency (Latham and Scully, 2002; Latham, 2006).

4.5. *Effective measures*

Zorach and Ulanowicz (2003) have presented effective measures for weighted networks.

(1) Effective connectivity:

$$C_z = \prod_{i,j=1}^{n} (T_{ij}^2/(T_i.T._j))^{-(0.5T_{ij}/T..)}$$

Effective connectivity is similar to link density in unweighted networks. However it has weight for the size of links.

(2) Effective flows:

$$F_z = \prod_{i,j=1}^{n} (T_{ij}/T..)^{-T_{ij}/T..}$$

Effective flows are similar to the number of links in unweighted networks.

(3) Effective nodes:

$$N_z = \prod_{i,j=1}^{n} (T^2/(T_i.T._j))^{(0.5T_{ij}/T..)}$$

Effective nodes are the weighted mean of the normalized throughflow of each node.

(4) Effective roles:

$$R_z = \prod_{i,j=1}^{n} (T_{ij}T../(T_i.T._j))^{T_{ij}/T..}$$

Effective roles is a measure of weighted, differentiated, distinct functions in a network, i.e., sectors that uniquely take input and pass matter/energy to one destination (Latham, 2006).

4.6. *Cycling index*

Cycling index was firstly proposed by Finn (1976, 1978, 1980) and Patten *et al.* (1976). It is a matrix procedure for calculating the proportion of cycled throughflow in a system. There are three versions of cycling indices. One of the versions is:

$$FCI = TST_c/TST$$

where TST: total network compartmental throughflows.; TST_c total system cycled throughflow (Latham, 2006).

4.7. *Homogenization index*

Homogenization index was developed by Fath and Patten (1999a) to measure network homogenization. It measures the evenness of flow in a network. Homogenization degree is determined by comparing the evenness of flow in a matrix accounting cycling and direct flow against the normalized direct flow matrix (Latham, 2006). Cycling generally increases the evenness of flow in a network; cycling and homogenization are thus related (Fath and Patten, 1999a):

$$CV_N = \left[\sum_{i,j=1}^{n} (m^N - m_{ij}^N)^2 / (n^2 - 1) \right]^{0.5} \bigg/ m^N$$

$$CV_G = \left[\sum_{i,j=1}^{n} (m^G - m_{ij}^G)^2 / (n^2 - 1) \right]^{0.5} \bigg/ m^G$$

where CV_N: homogenization for the integral matrix (including cycled flow); CV_G: homogenization for the direct flow matrix; m^N, m^G: average element from the integral flow matrix N and direct flow matrix G, respectively; m_{ij}^N, m_{ij}^G: elements from the integral flow matrix N and direct flow matrix G, respectively.

A lower CV means an even flow in the matrix. In addition, homogenization occurs if $CV_G > CV_N$ (Latham, 2006).

4.8. *Dominance index of indirect effects*

Higashi and Patten (1986, 1989) and Fath and Patten (1999b) presented an index to describe the dominance of indirect effects:

$$i/d = \sum_{i,j=1}^{n} (n_{ij} - \delta_{ij} - g_{ij}) \bigg/ \sum_{i,j=1}^{n} g_{ij}$$

The denominator denotes the direct processes and the numerator of right term denotes the indirect processes. The dominance degree of indirect processes increases with the index value.

4.9. *Amplification matrix and utility matrix*

Amplification matrix is a binary matrix to describe which links are providing more energy/matter than they are directly capable of producing (Fath and Patten, 1999b; Latham, 2006). In this matrix, elements are of value one where any element of the diagonal in the $(I - G)^{-1}$ matrix is greater than one. These are the links where amplification occurs.

Both utility matrix and synergism (b/c) are used to describe the degree of positive interactions in the network (Fath and Patten, 1999b). Utility matrix is expressed as:

$$(I - D_P)^{-1}$$

where $D_P = (d_{ij})$ is non-dimensional direct flow-based utility matrix, $d_{ij} = (T_{ij} - T_{ji})/T_i$. The positive elements of the matrix denote positive interactions (mutualism) and negative elements denote negative interactions (competition). The b/c index is the ratio of positive to negative interactions.

5. Software for Network Analysis

Several software packages are available for network analysis. They are basically free of charge.

5.1. *Netwrk*

Netwrk can be used to calculate input-output structure matrices, trophic chain and tropic aggregation, and ascendency metrics, and to identify biogeochemical cycles (Ulanowicz, 2004; Latham, 2006). Netwrk was designed to run under MS-DOS.

Netwrk and documentation is available at (Ulanowicz and Kay, 1991) http://www.cbl.umces.edu/~ulan/ntwk/network.html.

5.2. *EcoNetwrk*

EcoNetwrk is a Microsoft Windows-based version of Netwrk 4.2 (Ulanowicz, 2004). It can be used to conduct the following analyses: (1) input/output analysis; (2) the determination of trophic status and the identification of an equivalent linear food chain; (3) analysis of biogeochemical cycling and the supporting flows; (4) calculation of ecosystem indices.

EcoNetwrk and documentation is available at http://www.glerl.noaa.gov/EcoNetwrk/.

The input files of EcoNetwrk include the following: (1) **ASCI files**. These files preserve biomasses, imports, exports, respiration, diet exchanges, consumption, production, ratios, assimilation efficiencies, egestion, diet proportions, and import proportions; (2) **SCOR files.** They preserve biomasses, imports, exports, respiration, and diet exchanges. Ratios and variable states are not saved; (3) **CSV files.** They are comma separated value files that are easily imported into Excel. They preserve the same values as SCOR files. However, Excel should only be used to change cell values. EcoNetwrk will not correctly interpret the file if some columns of an input file are deleted or rearranged.

As demonstrated in EcoNetwrk documentation, the summary page (Fig. 1) of EcoNetwrk indicates the existing state of the network. Values of this page can be changed. Biomass and ratios page contains biomass and ratio information. The ratios listed on this page are C/B (Consumption/Biomass), P/B (Production/Biomass), R/B (Respiration/Biomass), and R/C

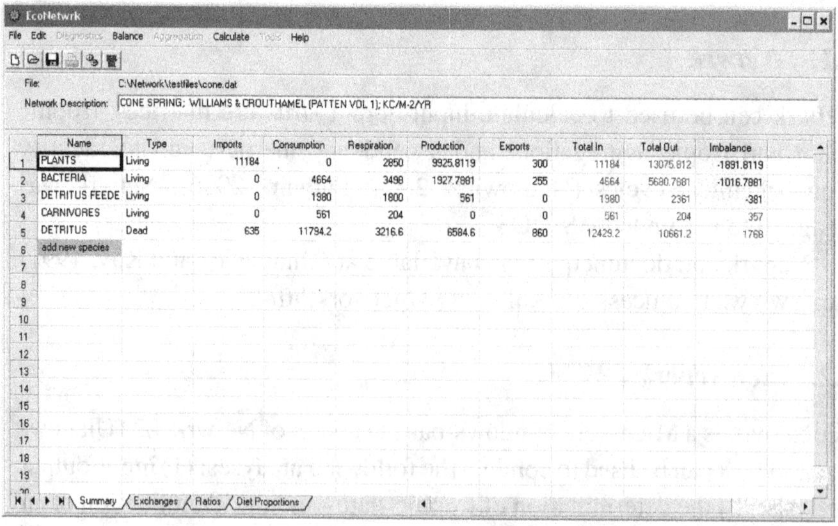
EcoNetwrk
File Edit Diagnostics Balance Aggregation Calculate Tools Help
File: C:\Network\testfiles\cone.dat
Network Description: CONE SPRING; WILLIAMS & CROUTHAMEL (PATTEN VOL 1); KC/M-2/YR
Name Type Imports Consumption Respiration Production Exports Total In Total Out Imbalance
1 PLANTS Living 11184 0 2850 9925.8119 300 11184 13075.812 -1891.8119
2 BACTERIA Living 0 4664 3498 1927.7881 255 4664 5680.7881 -1016.7881
3 DETRITUS FEEDE Living 0 1980 1800 561 0 1980 2361 -381
4 CARNIVORES Living 0 561 204 0 0 561 204 357
5 DETRITUS Dead 635 11794.2 3216.6 6584.6 860 12429.2 10661.2 1768
6 add new species
Summary Exchanges Ratios Diet Proportions

Figure 1. Summary page of EcoNetwrk.

(Respiration/Consumption). Diet proportions page summarizes the relative components of each predator's diet. Imports are included which makes the divisor Total Inputs (Consumption + Imports). The totals must sum to one before network analysis can proceed. The exchange pages contain such information as "who eats how much of whom?" and "who eats me?", etc.

5.3. *Ecopath*

Ecopath can be used to calculate ascendency measures (Ulanowicz, 1980, 1986, 1997), a decomposition of network cycles (Ulanowicz, 1983, 1986), mixed trophic impact (Ulanowicz and Puccia, 1990), and trophic aggregation (Ulanowicz, 1995), etc.

5.4. *Pajek*

Pajek (Fig. 2) is used to conduct analysis and visualization of large scale networks (Batagelj and Mrvar, 2010). It supports abstraction by recursive factorization of a large network into smaller networks, which can be treated

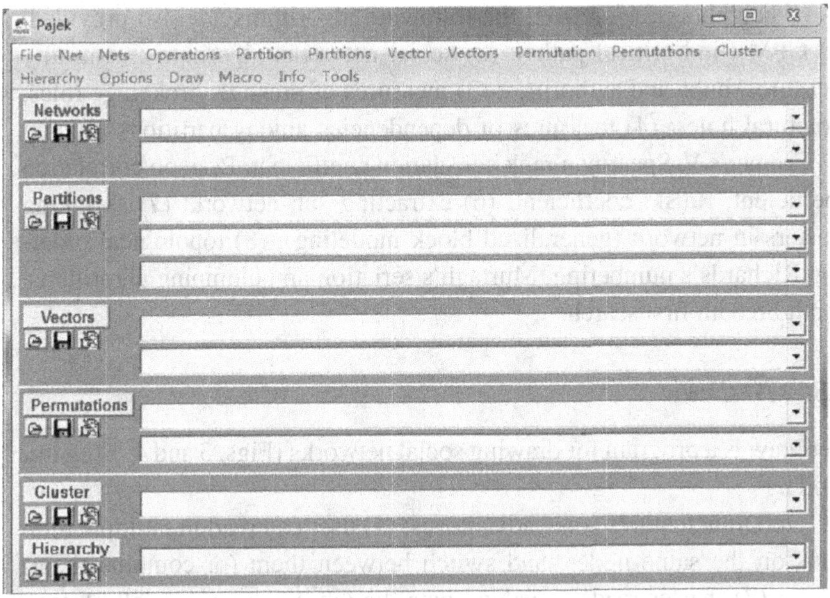

Figure 2. Pajek.

using more sophisticated methods. Pajek provides some powerful visualization tools and implements a selection of efficient algorithms for analyses of large networks.

Pajek uses several data structures as follows:

(1) **Network.** the main object, which includes vertices and lines.
(2) **Permutation.** Reordering of vertices.
(3) **Vector.** A set of values of vertices.
(4) **Cluster.** A subset of vertices.
(5) **Partition.** Partition shows for each vertex to which cluster the vertex belongs.
(6) **Hierarchy.** Hierarchically ordered clusters and vertices.

Pajek can be used to conduct simplifications and transformations like deleting loops, multiple edges, transforming arcs to edges, etc., calculate components (strong, weak, biconnected, and symmetric components), make decompositions (symmetric-acyclic, hierarchical clustering), find paths (shortest path(s), all paths between two vertices), calculate flows (maximum flow between two vertices), and make neighborhood analysis (k-neighbors). Moreover, the following algorithms are also provided: (1) CPM (find critical paths); (2) social networks algorithms: centrality measures, hubs and authorities; (3) measures of prestige, brokerage roles, structural holes; (4) measures of dependencies among partitions/vectors; (5) Cramer's V, Spearman rank correlation coefficient, Pearson correlation coefficient, Rajski coefficient; (6) extracting sub-network; (7) shrinking clusters in network (generalized block modeling); (8) topological ordering, Richards's numbering, Murtagh's seriation and clumping algorithms, depth/breadth first search.

5.5.   *NetDraw*

NetDraw is a program for drawing social networks (Figs. 3 and 4; Borgatti, 2011).

NetDraw features some advantages: (1) user can read in multiple relations on the same nodes, and switch between them (or combine them) easily; (2) if user reads in valued data, he (she) can sequentially "step" through different levels of dichotomization, selecting only strong ties, only

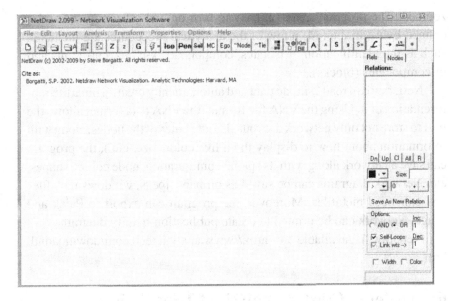

Figure 3. NetDraw.

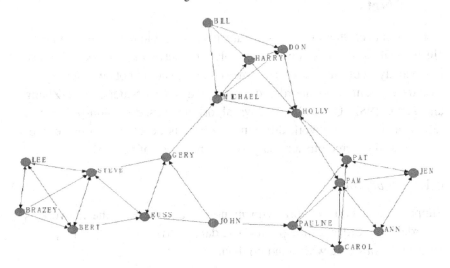

Figure 4. A network drawn by NetDraw.

weak ties, etc. User can choose to let the thickness of lines correspond to strength of ties; (3) the program makes it convenient to read in multiple node attributes for use in setting colors and sizes of nodes, as well as rims, labels, etc. Diagram can be rotated, flipped, shifted, resized and zoomed. The program makes it easy to turn on and off groups of nodes defined by a variable; (4) a few of the analytical procedures are included, such as the identification of isolates, components, k-cores, cut-points and bi-components (blocks).

NetDraw can read 2-mode data and automatically create a bipartite representation of it. Using the VNA file format (the VNA data format allows the user to store not only network data but also attributes of the nodes, along with information about how to display them like color, size, etc.), the program can save a network along with its spatial configuration, node colors, shapes, etc. Network diagrams can be saved as bitmaps, jpegs, windows metafiles and enhanced metafiles. Moreover, the program can export to Pajek and Mage. Network can be printed to create publication-quality diagrams.

NetDraw is available at: http://www.analytictech.com/downloadnd.htm.

6. Ordered Cluster Analysis of Time Series of Network

Conventional cluster analysis is not always reasonable if we want to make a cluster analysis to switch on, for example, time series of network dynamics. In the analysis of time series, the time sequence should not be upset. In this case the one-dimensional ordered cluster analysis is a better choice (Zhang and Fang, 1982; Qi, 2005). A Java algorithm, based on non-parametric test p of between-network difference, was thus developed to make one-dimensional ordered cluster analysis on time series of network.

6.1. *Algorithm*

Suppose there is a network with m nodes, and state of the network is recorded on n ordered time points. The data matrix is $(x_{ij})_{m*n}$. First, standardize the raw data with standard deviation:

$$a_{ij} = (x_{ij} - x_{bi})/s_i,$$

where,

$$x_{bi} = \sum_{j=1,...,n} x_{ij}/n,$$

$$s_i = \left(\sum_{j=1,\ldots,n} (x_{ij} - x_{bi})^2 / (n-1) \right)^{1/2}, \quad i = 1, 2, \ldots, m.$$

The second standardization measure is the range:

$$a_{ij} = (x_{ij} - \min x_{ik}) / (\max x_{ik} - \min x_{ik}),$$
$$i = 1, 2, \ldots, m; \quad j = 1, 2, \ldots, n.$$

where $\max x_{ik} = \max(x_{i1}, x_{i2}, \ldots, x_{in})$, $\min x_{ik} = \min(x_{i1}, x_{i2}, \ldots, x_{in})$. The third measure is a ratio:

$$a_{ij} = x_{ij} / \max x_{ik},$$
$$i = 1, 2, \ldots, m; \quad j = 1, 2, \ldots, n.$$

Calculate between-time Euclidean distance or Manhattan distance or Pearson correlation coefficient based distance, or Jaccard coefficient:

$$d_{i\,i+1} = \left(\sum_{k=1,\ldots,m} (a_{ki} - a_{k\,i+1})^2 / m \right)^{1/2}$$

$$d_{i\,i+1} = \sum_{k=1,\ldots,m} |a_{ki} - a_{k\,i+1}| / m$$

$$d_{i\,i+1} = 1 \sum_{k=1,\ldots,m} (a_{ki} - a_{ibar})(a_{k\,i+1} - a_{i+1bar}) \Big/$$

$$\left(\sum_{k=1,\ldots,m} (a_{ki} - a_{ibar})^2 \sum_{k=1,\ldots,m} (a_{k\,i+1} - a_{i+1bar})^2 \right)^{1/2}$$

$$d_{i\,i+1} = (b_i + b_{i+1}) / (c_i + c_{i+1} - e)$$
$$i = 1, 2, \ldots, n-1$$

where b_i is the non-zero number present at time i but not at time $i+1$, b_{i+1} is the non-zero number present at time $i+1$ but not at time i, c_i and c_{i+1} is the non-zero number at time i and time $i+1$ respectively, and e is non-zero number shared by time i and time $i+1$.

Using the non-parametric randomization statistic test (Zhang *et al.*, 2006; Zhang, 2011b), statistic p value for between-time difference may be calculated. A smaller p value means a greater difference between two time points. Thus take $r = 1 - p$ as between-time distance, and the distance between time points i and $i + 1$ is $r_{i\,i+1} = 1 - p_{i\,i+1}$.

Let $r_{i\,i+1} = 1 - p_{i\,i+1}$, search for the smallest distance $r_{i\,i+1}$, $i = 1, 2, \ldots, n-1$ and combine the two corresponding times i and $i+1$ into the same cluster. Similarly, take the distance between two most adjacent times belonging to the two adjacent clusters as the cluster distance, search for the smallest cluster distance and combine the clusters into the same cluster. Finally, all the above times will be combined into one cluster (Zhang and Fang, 1982). The smallest distance in each clustering is the cluster distance.

If between-cluster distance, $r = 1 - p$, is greater than the significance degree, i.e., 0.99, or 0.95, then the two clusters are significantly different.

The algorithm, NetOneDimCluster, is implemented as a Java program based on JDK 1.1.8, in which several classes and an HTML file is included.

6.2. *Application*

There is an ecosystem network with six compartments (nodes), and biomass of each compartment in the network changes with time (Fig. 5). The state of network is recorded seven times as the change of time (Table 1).

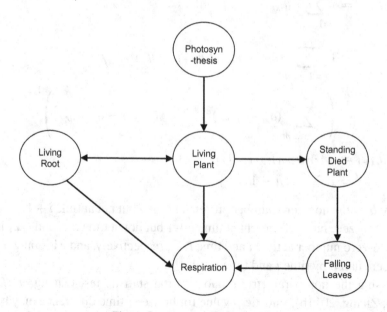

Figure 5.   An ecosystem network.

Table 1. Ecosystem dynamics for some time points.

t	1	2	3	4	28	29	30
Photosynthesis	200	200	200	200	201	201	201
Living plant	216	414	617	812	5115	5293	5407
Living root	78	80	77	81	454	470	547
Standing died plant	2	2	2	2	26	28	30
Falling leaves	12	16	17	18	18	18	18
Respiration	50	46	45	45	149	152	161

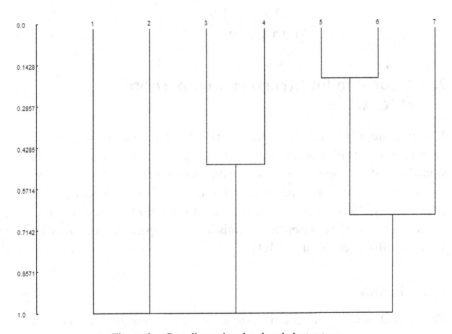

Figure 6. One-dimensional ordered cluster tree.

Using the algorithm above (Euclidean distance) and the raw data in Table 1, the one-dimensional ordered cluster tree can be obtained, as indicated in the following and Fig. 6, where numbers 1–7 represents time points 1, 2, 3, 4, 28, 29, and 30:

$$r = 1 - p = 0.0:$$
$$(1)\ (2)\ (3)\ (4)\ (5)\ (6)\ (7)$$

$$r = 1 - p = 0.018:$$

(1) (2) (3) (4) (5 6) (7)

$$r = 1 - p = 0.1909:$$

(1) (2) (3 4) (5 6) (7)

$$r = 1 - p = 0.486:$$

(1) (2) (3 4) (5 6 7)

$$r = 1 - p = 0.663:$$

(1) (2 3 4) (5 6 7)

$$r = 1 - p = 1.0:$$

(1 2 3 4 5 6 7)

7. Algorithm for Structure Comparison of Networks

The structure of network refers to the node degree, network connectance, aggregation strength (Dormann, 2011; Zhang, 2011a,b,c), etc. We occasionally need to compare the structure difference between networks. Nonparametric statistics may be used in the difference comparison (Solow, 1993; Manly, 1997; Zhang, 2007a,b). A Java algorithm, based on previous studies, was thus presented to statistically compare between-network structure difference (Zhang, 2011c).

7.1. *Algorithm*

The algorithm is used to compare the difference in structure composition between two networks.

Suppose that a_{ij} is the mass (or degree, etc.) of node j in network i, $i = 1, 2, \ldots, n$; $j = 1, 2, \ldots, s$. First, define between-network distance measures, i.e., Euclidean distances, Manhattan distances, Chebyshov distance, Pearson correlation (based distance), are as follows:

$$d_{ij} = \left(\sum_{k=1}^{s} (a_{ik} - a_{jk})^2 / s \right)^{0.5}$$

$$d_{ij} = \sum_{k=1}^{s} |a_{ik} - a_{jk}|/s$$

$$d_{ij} = \max_{k} |a_{ik} - a_{jk}|$$

$$d_{ij} = 1 - \sum_{k=1}^{s} ((a_{ik} - a_{ibar})(a_{jk} - a_{jbar})) \Bigg/$$

$$\left(\sum_{k=1}^{s} (a_{ik} - a_{ibar})^2 \sum_{k=1}^{s} (a_{jk} - a_{jbar})^2 \right)^{0.5}$$

where a_{ibar} and a_{jbar} are means of a_{ik}'s and a_{jk}'s.

If $\min a_{ij} < 0$, then let $a_{ij} = a_{ij} - \min a_{ij}$, $i = 1, 2, \ldots, n$; $j = 1, 2, \ldots, s$. Suppose z_{ij} is the decimal numbers of a_{ij} if network data contains the decimal value a_{ij}, then calculate $c_{ij} = 10^{z_{ij}}$. Let $a_{ij} = a_{ij} \max c_{kl}$, $i = 1, 2, \ldots, n$; $j = 1, 2, \ldots, s$. Through these transformations all of the values in network data become integers which are equivalent to numbers of individuals. If no difference exists, then the distribution of individuals in networks i and j will be a result of allocating the mixed network values at random into two networks of size equal to those of the original network (Solow, 1993; Manly, 1997; Zhang, 2007a,b). Assume the two networks to be tested are i and j, which contain $\sum_{k=1}^{s} a_{ik}$ and $\sum_{k=1}^{s} a_{jk}$ individuals respectively. The $\sum_{k=1}^{s} a_{ik} + \sum_{k=1}^{s} a_{jk}$ individuals of the combined network are randomly reallocated into two randomized networks with $\sum_{k=1}^{s} a_{ik}$ and $\sum_{k=1}^{s} a_{jk}$ labeled individuals. Calculate the expected absolute distance between the two randomized networks and compare whether it is not less than the absolute distance between the true networks i and j. Repeat the simulation many times, calculate whether the number of the expected are not less than the absolute distance between i and j, and take the percentage as the p value. The p value is used to make statistical test. The threshold p value for test may be defined as 0.05, 0.01, etc. If the calculated p value is less than p threshold, then the structure composition of networks i and j are statistically different.

The algorithm, NetStructComp, is implemented as a Java program based on JDK 1.1.8, in which several classes and an HTML file is included. In network data file, the first row is ID numbers of nodes and the first column is ID numbers of networks.

7.2. Application

Let us choose the weed data of rice fields in four cities (networks) of Pearl River Delta, China. In total 25 plant families (nodes) were found (Wei, 2010), as indicated in Table 2.

Choose Euclidean distance measure, significance level $p = 0.01$, and 1000 randomizations, the results are as follows:

Table 2.   Abundance of plants around rice fields in four cities of China.

Plant family	Zhongshan	Zhuhai	Dongguan	Guangzhou
Gramineae	1056.9	184.6	439.3	193.6
Compositae	11.1	95	43.3	63.4
Amaranthaceae	31.1	56	93.4	49
Commelinaceae	0	52.2	14.4	1.4
Onagraceae	0	0.1	2.3	1
Urticaceae	0.3	0	11.9	3.4
Menispermaceae	0	0	0	0.1
Cyperaceae	0	0	0	26.1
Caryophyllaceae	0	0	5.3	6.7
Polygonaceae	0.4	4.2	6.9	3.8
Acanthaceae	0	0	0	0.3
Solanaceae	0.1	0	0.2	0.4
Umbelliferae	0	34.9	0	4.8
Lythraceae	0	0	0	1.6
Scrophulariaceae	0.7	3.9	1.7	2.4
Oxalidaceae	0	0	0	0.2
Chenopodiaceae	1.1	0.4	0.3	0.1
Haloragaceae	0	0	0	4.6
Campanulaceae	0	0	0	0.7
Plantaginaceae	0	0.3	0	0
Rubiaceae	0	0.1	0	0
Euphorbiaceae	0	0	0.1	0
Convolvulaceae	0	0.1	0	0
Pontederiaceae	0	0.1	0	0
Portulacaceae	24.6	0	8	0

Network pairs with significant statistic difference in structure (with *p* values):

$$(1, 2)(0.0) \quad (1, 3)(0.0) \quad (1, 4)(0.0)$$

$$(2, 3)(0.0) \quad (2, 4)(0.0)$$

$$(3, 4)(0.0)$$

It is obvious that all network pairs have significant statistic difference.

8. Test Samples' Homogeneity and Examine Sampling Completeness

In food web sampling studies, we need to record all possible taxa (species, families, etc.) in the community. Enough samples should be taken to record enough taxa. To examine sampling completeness, a yield-effort curve may be drawn, which plots the cumulative number of taxa caught or observed (y-axis) against the cumulative effort of sampling (x-axis) (Cohen, 1978; Dickerson and Robinson, 1985; Cohen *et al.*, 1993; Zhang and Schoenly, 1999). If sampling stops while the yield-effort curve is still rapidly increasing, then the community derived from this sampling is incomplete. Nevertheless, if sampling ceases when the slope of the yield-effort curve reaches zero or close to zero, the sampling is probably complete. In addition, the correlation based eco-interaction network studies require the homogeneity of samples or environment. To ensure sample homogeneity is also a necessity work in these studies. Bias from sample order can be corrected by bootstrap procedure. However, variation in curve shape due to environmental heterogeneity remains a likely significant source of sampling error (Zhang and Schoenly, 1999). Coleman *et al.* (1982) developed a statistical model to test whether individuals among the samples (of definable size) obey the random placement hypothesis which assumes a lack of correlation in the location of individuals (Zhang and Schoenly, 1999). The model of Coleman *et al.* can test sample homogeneity and examine sampling completeness. A Java program, based on the model of Coleman *et al.* (1982) and the algorithm of Zhang and Schoenly (1999) was described here. Compared to the algorithm of Zhang and Schoenly (1999), it gives the conclusion for completeness of sampling and can be easily run on web browser.

8.1. *Algorithm*

Under the random placement hypothesis, consider a collection C of N individuals from S taxa, with ni individuals in C belonging to the ith taxon, and suppose that each member of C occurs in one of k non-overlapping samples that have areas $a_1, a_2, \ldots, a_k$. The number of taxa, s, in a given region is a random variable whose magnitude depends on the area a of the region, and the relative area is defined as $\alpha = a / \sum a_i$. The mean number of taxa, s, and the variance σ^2 are calculated as follows:

$$s(\alpha) = S - \sum (1 - \alpha)^{ni},$$

$$\sigma^2(\alpha) = \sum (1 - \alpha)^{ni} - \sum (1 - \alpha)^{2ni}.$$

The method to test sample homogeneity is to compare the observed mean taxa richness *vs.* the sample size with the expected taxa richness *vs.* the sample size curve (Zhang and Schoenly, 1999). If 95% of the plotted points (means) of the observed curve fall into two standard deviations outside the expected curve, then the observed samples are statistically more heterogeneous in taxa composition (at the 0.05 level) than sampling error (alone) can account for (Coleman *et al.*, 1982). Thus we can conclude that these samples are more heterogeneous in taxonomic composition than is expected under the random placement hypothesis.

Bootstrap procedures are used to produce the taxa richness *vs.* the sample size curves. The curves plot the cumulative number of taxa, defined as the sum of the number of taxa in the previous sample(s) and the number of taxa in the present sample that were not observed in any previous sample. For the first sample, the cumulative number of taxa is defined to equal the number of taxa found in this sample.

If random placement hypothesis is met, the samples are homogeneous, or else they are heterogeneous. If the difference of the number of taxa between the last two (cumulative) sample sizes is less than desired percent threshold, then most of the taxa are considered to be recorded and the sample size is enough.

The algorithm is implemented as a Java program, SampHomoTest, based on JDK 1.1.8, in which several classes and an HTML file is included. In sampling data file, the first row is sample ID numbers and the first column is taxon ID numbers.

8.2. Application

We obtained a set of arthropod data investigated in rice fields of Guangzhou, China (Data set 1: 35 samples; 19 families; Data set 2: 54 samples; 23 families; Data set 3: 60 samples; 23 families; Data set 4: 60 samples; 27 families), investigated in 2006 (Zhou, 2007).

The results from the algorithm above showed that four arthropod communities are all environmentally homogeneous (all observed points fell inside the confidence interval) and the sampling is complete for the four arthropod communities. The results for a data set (35 samples, 19 families) are listed in Table 3.

Table 3.   Test results for a data set.

Sample size	Mean observed number of taxa (ONT)	Expected number of taxa (ENT)	Standard deviation of expected number of taxa	Lower limit of ENT	Upper limit of ENT	Lower limit of ONT	Upper limit of ONT
1	5.57	6.799	1.311	4.175	9.423	1.458	8.541
2	7.743	8.561	1.464	5.632	11.49	3.533	10.466
3	9.313	9.812	1.534	6.743	12.882	5.598	12.401
4	10.428	10.81	1.557	7.694	13.926	6.676	13.323
5	11.536	11.637	1.555	8.526	14.748	7.737	14.262
6	12.208	12.338	1.538	9.261	15.416	8.887	15.112
7	12.935	12.942	1.514	9.913	15.972	8.991	15.008
8	13.461	13.469	1.486	10.496	16.442	10.105	15.894
9	13.984	13.933	1.456	11.02	16.847	9.994	16.005
10	14.41	14.346	1.426	11.493	17.199	11.274	16.725
11	14.688	14.716	1.395	11.924	17.508	11.199	16.8
12	15.152	15.051	1.365	12.319	17.782	12.226	17.773
13	15.455	15.355	1.336	12.682	18.028	12.282	17.717
14	15.673	15.634	1.307	13.019	18.248	12.27	17.729
15	16.001	15.89	1.278	13.333	18.447	13.212	18.787
16	16.233	16.128	1.249	13.628	18.628	13.352	18.647
17	16.444	16.35	1.221	13.907	18.793	13.428	18.571
18	16.638	16.557	1.192	14.172	18.943	13.441	18.558
19	16.789	16.753	1.163	14.425	19.08	13.413	18.586

(*Continued*)

Table 3.   (*Continued*)

Sample size	Mean observed number of taxa (ONT)	Expected number of taxa (ENT)	Standard deviation of expected number of taxa	Lower limit of ENT	Upper limit of ENT	Lower limit of ONT	Upper limit of ONT
20	16.908	16.937	1.133	14.669	19.204	13.444	18.555
21	17.142	17.112	1.103	14.906	19.318	14.574	19.425
22	17.326	17.278	1.071	15.136	19.42	14.619	19.38
23	17.504	17.437	1.037	15.362	19.512	14.681	19.318
24	17.605	17.59	1.001	15.586	19.593	14.717	19.282
25	17.793	17.736	0.964	15.808	19.664	14.771	19.228
26	17.895	17.878	0.923	16.031	19.724	14.914	19.085
27	18.055	18.015	0.879	16.257	19.773	15.932	20.067
28	18.186	18.148	0.83	16.487	19.809	16.048	19.951
29	18.253	18.277	0.776	16.724	19.831	16.121	19.878
30	18.432	18.404	0.716	16.971	19.837	16.375	19.624
31	18.559	18.527	0.647	17.232	19.823	16.48	19.519
32	18.67	18.649	0.567	17.515	19.783	16.72	19.279
33	18.805	18.767	0.467	17.832	19.703	16.962	19.037
34	18.9	18.884	0.334	18.215	19.553	17.24	18.759
35	19	19	0	19	19	19	19

PART III

Agent-based Modeling

Agent-based Modeling

1. Complex System

Complex systems have the following properties (Li *et al.*, 2007):

(1) **Emergency.** Emergency is a dynamic process from lower level to higher level, from locality to globe, and from micro level to macro level. It stresses that the interactions between individuals result in different functions and properties and behaviors from that of individuals, which leads to a system with certain functional characteristics and purposeful behaviors that differ by the nature of individuals.

(2) **Nonlinearity.** Usually, complex systems exhibit nonlinear behaviors. The global behaviors of a complex system cannot be derived from individuals' behaviors.

Suppose a system is represented by the model:

$$y = f(x).$$

If the following condition is satisfied:

$$f(\alpha x_1 + \beta x_2) = \alpha f(x_1) + \beta f(x_2),$$

where $\alpha, \beta \in R$, then it is a linear system, otherwise it is a nonlinear system (Zhang, 2007b).

(3) **Modularity/Hierarchy.** Logistically, a complex system is an aggregation of interactive and correlative individuals/objects (modularity). Meanwhile, a complex system has a hierarchical structure (hierarchy). The objects at a hierarchical level aggregate to form the objects at its parent hierarchical level.

(4) **Information flow/Associativity.** In a complex system, there are large amounts of information/matter interactions between individuals/objects. Changes of any individual/object might affect other individuals/objects, or even the entire system (Harman *et al.*, 2009).

(5) **Dynamicness.** The complex system will change with the time.

(6) **Incomputability.** The incomputability of complex systems includes two aspects: (A) Behaviors of a complex systems cannot be described by using deduction, induction, or other formalization methods. (B) Process of a complex system can be approximated with the inference system based on rules.

The study and description of complex systems uses differential equations, artificial neural networks, agent-based modeling, network analysis, etc.

Differential equations can be used for modeling complex systems. The model is established in a top-down way (Hraber and Milne, 1997). These models are always differential or finite difference equations on one or more species (Zhang and Gu, 2001). In these differential equations, the types of interactions should be predefined. Pairwise interaction coefficients represent the interactions between species. All species are assumed to interact with equal probability, which is a characteristic of interactive system with fully mixed or mean field. In a sense, a differential equation model is the agent-based model of uniform field. In differential equations, we usually reflect spatial heterogeneity of interactions by defining adjacent interactions and including local abiotic factors. These models are used to predict which species will dominate community as the change of time. The parameters of differential equations may approximate the apparent properties of a biological population but the parameters in differential equation models approximate phenomenological properties of populations without identifying mechanisms or considering variation among individuals and, thus are hard to reflect complex interactions and to include various behaviors simulated. In addition, the complexity and incomputability of differential equation models will exponentially grow as the rise of system's complexity. Hence they can only be used to describe systems with lower complexity.

Artificial neural networks (ANNs) are used to handle highly complex and nonlinear ecological problems (Zhang and Barrion, 2006; Zhang, 2007a, b; Zhang *et al.*, 2007, 2008a, b; Zhang and Zhang, 2008; Zhang and

Wei, 2009; Zhang, 2010). Through learning from samples, ANNs store the intrinsic mechanism of the system or data sets studied as connected weights of network. In a sense ANNs are models that lie between emphirical models and mechanistic models. There are many types of ANNs. Their capability and effectiveness are depended on three factors:

(1) Complexity of neuron models.
(2) Speed and efficiency of leaning rules.
(3) Topological structure of network.

The first ANN model of importance is Multi-layer Perceptron (MLP). MLP is a feed-forward ANN that the neurons distribute in every layer. The neurons at each layer only connect to the neurons in posterior layer and thus loops are not allowed. A standard setting for feed-forward ANNs is that there are three types of layers, i.e., input layer, hidden layer, and output layer. There are currently a lot of software packages on ANNs among which most of them tend to establish feed-forward ANNs that is used in classification, optimization, and regression. ANNS, like BP, radial basis function networks, self-organizing networks, etc., have been widely used in ecology including community succession (Zhang and Barrion, 2006; Zhang *et al.*, 2007, 2008a, b; Zhang, 2010).

ANNs are learning models which need enough sample data to train and learn from. In a sense, ANNs have the characteristic of fully mixed or mean field systems.

Two powerful tools to describe complex systems are agent-based modeling (ABM) and network analysis. As discussed in the preceding chapters, network analysis is based on graph theory, etc. It treats individuals/objects as nodes in the network, which allows people to analyze the large scale structure of a complex system. This methodology makes it easy to examine the effects of interactions, characterize important elements, and describe global structure of the system. Nevertheless, this methodology is in general insufficient to describe system dynamics.

2. Agent-based Modeling

Agent-based modeling (ABM) is, in a sense, equivalent to individual-based modeling. The thinking of ABM was originally from complex adaptive system (CAS). A complex adaptive system can spontaneously organize itself and dynamically reconstruct its components in order to survive in the environment. Holland (1995) held that a complex adaptive system has the following properties:

(1) **Aggregation.** This allows formation of a population.
(2) **Nonlinearity.** Simple extrapolation is invalid.
(3) **Flows.** Flows allow the transition and transformation of resources and information.
(4) **Diversity.** Different agents may have different behaviors, which usually results in a robustness of the system.

The mechanisms of complex adaptive system are:

(1) **Labels.** This allows agents to be defined and identified.
(2) **Internal pattern.** This allows agents to make inference to their world.
(3) **Establishment of blocks.** This allows components and entire system being constructed from simple components at various hierarchies.

These properties and mechanisms provide ABM with a useful frame.

Earlier ABM was originated from cell automata (CA) (Gardner, 1970). In answering the question: if a machine can be programmed to reproduce itself, proposed by Von Neumann, the physicist Stanislaw Ulam proved being true by using CA.

A typical CA is a two-dimensional mesh or lattice or dot matrix composed of cells (grids). At any time and point, each cell has a finite number of states. A set of simple rules are used in previous states for determining current value of a cell. The next value of a cell is dependent upon its current value and the values of its eight adjacent cells. Updating rules of each cell are the same.

Wolfram found that the rules will yield astonishing spontaneous patterns in CA (Wolfram, 2002). These patterns directly correspond to extensive algorithms and logic systems. He thought that simple rules will result in a complexity similar to the true world.

ABM is a bottom-up modeling method. It is used to model the complex systems containing spontaneous and interactive agents (Topping *et al.*, 2003; Qi *et al.*, 2004; Li and Ma, 2006; Chen *et al.*, 2008), which make it a powerful tool for analyzing global behaviors of complex systems.

ABM models adaptive system dynamics based on adaptation mechanism of individuals. It is thoroughly different from differential equation modeling. ABM is even considered as the third scientific methodology besides induction and deduction. ABM is related to a lot of sciences, including complexity science, systems science, computer science, etc.

As the systems studied becomes more and more complex, conventional simulation tools exhibit obvious shortages. ABM will play important roles in these aspects: (1) Conventional methods have reached their limitations. (2) Data are organized as databases. Data can be stored hierarchically in databases. (3) Computational capabilities have been largely enhanced and allow extensive simulation.

ABM was originally used to propose hypothesis on the behaviors and mechanism of a system, to explain system phenomena, and now further to propose management strategies. It has been successfully used in organizational simulation (risks, organizational design, molecular self-organizing), diffusion simulation (diffusion dynamics), flow simulation (traffic, flow management), market simulation (stock market), etc (Bonabeau, 2002). For example, traditional risk analysis appears to be unrelated to ABM. However, risk can be in some cases treated as a spontaneous property of a system. ABM is thus available for solving these problems.

ABM can be used in these cases (Bonabeau, 2002):

(1) Interactions between agents are complex, nonlinear, and discrete.
(2) Spatial factors are very important and the locations of agents are not fixed.
(3) Population is heterogeneous. Each individual is different from others.
(4) Complex and heterogeneous interactions.
(5) Agents exhibit complex and diverse behaviors, including learning and adaptation.

ABM can be jointly used with other modeling techniques like systematic dynamics, etc. Statistics, like PCA, is also useful in ABM.

2.1. *Agent and behaviors*

Agent can be defined in various ways. Any independent component, such as software, model, individual, etc., can be considered as an agent (Bonabeau, 2002). The behavior of an independent component might be a primitive response and decision, or even a complex adaptive intelligence. In general, the behaviors of an agent must be self-adaptive and can learn from its changing environment (Mellouli *et al.*, 2003). Casti (1997) maintained that the behavior rules of an agent must include two parts, basic rules, and the high-leveled rules that govern basic rules (rule-changing rules). Basic rules define necessary responses to the environment, and rule-changing rules define adaptation. Jennings (2000) held that the basic characteristic of an agent is that it can independently make decisions.

Different from differential equations, combining new agents and new agent types into ABM is easy. At lower levels differential equations can be included to approximate system's behaviors. At higher levels the interactions between agents are allowed to define aggregation behaviors. Aggregation behaviors are spontaneous behaviors generated by between-agent or agent-environment interactions.

ABM allows an agent and entire system to be consisted of more simple agents at various levels. Unlike AI (artificial intelligence) agent, the environmental responses and influences of an agent in ABM are always not so complex.

Agents in ABM must satisfy these criteria (Macal and North, 2005):

(1) An agent is an independent and identifiable individual which possesses a set of attributes and rules that govern its behaviors and decision capability. An agent is self-contained and independent. It has a boundary through which people can easily discern between outside agent and inside agent or shared characteristic.
(2) Each agent locates in a certain spatial position and interacts with its adjacent agents. An agent has a set of protocols that govern its interactions with other agents, such as communication protocol, the capability to affect its environment, etc. Agent is able to identify and discern the characteristics of other agents.
(3) Agent is goal-directed. Agent behaves to realize some goals.
(4) Agent is independent, autonomous and self-guided. At least within a finite range, agent can independently operate in its environment.
(5) Agent is flexible. It is capable of accumulating experiences and learning from the environment and adjusting its behaviors. This requires some form of memory. Agent possesses some high-level rules to adjust its

low-level behavioral rules. It has diverse attributes and behavioral rules.

To define agents, exactly specify their behaviors, and reasonably represent interactions between agents, which are fundamental to ABM.

Once agents are defined, we should exactly define their behaviors. First, it is necessary to determine a theory on behaviors and existing behavioral theories can be used. Agent may use various behavioral models, including if-then rule and threshold rule.

Behaviors of agents are the base for an existing or supposed system. Knowledge engineering and participative simulation can be used in defining behaviors. Knowledge engineering includes a series of techniques collected for organizing experts' knowledge.

2.2. *Procedures of agent-based modeling*

Main procedures of ABM include the following:

(1) Determine various types of agents and define behaviors of agents.
(2) Identify relations between agents, and construct interaction types between agents.
(3) Determine the platforms and environments for ABM, and set the strategies for ABM.
(4) Acquire necessary data for ABM.
(5) Validate the patterns of agents' behaviors and system's behaviors.
(6) Run ABM model, and analyze the output from the standpoint of linking the micro-scale behaviors of the agents to the macro-scale behaviors of the system.

To design an ABM, the key is software design and model development. Development time spans several highly staggered phases (Macal and North, 2005). At the design phase, the structure and function of the model should be defined. At implementation phase, we should develop a model based on the designed plan. Model will be used during the practical operation. In practice these phases are usually repeated several times to form a more detailed model.

Grimma *et al.* (2006) has also proposed a standard protocol for describing ABM. The core of the protocol is to structure the information about ABM in the same sequence. This sequence consists of seven elements, which can be grouped in three blocks: overview, design concepts, and details (Grimma *et al.*, 2006):

(1) The overview consists of three elements including purpose, state variables and scales, process overview and scheduling. It provides an overview of the overall purpose and structure of the model. It includes the declaration of all objects (classes) describing the models entities (different types of individuals or environments) and the scheduling of the model's processes.

(2) The design concepts describe the general concepts underlying the design of the model. The purpose of this element is to link model design to general concepts identified in the field of complex systems. These concepts include questions about emergence, the type interactions among individuals (agents), whether individuals consider predictions about future conditions, or why and how stochasticity is considered.

(3) The details include three elements, i.e., initialization, input, and sub-models, which present the details that were omitted in the overview. The sub-models implementing the model's processes are particularly described in detail. All information required to completely re-implement the model and run the baseline simulations should be provided here.

The logic behind the protocol sequence is: context and general information is provided first (overview), followed by more strategic considerations (design concepts), and finally more technical details (details).

2.3. *Tools and platforms for agent-based modeling*

Despite ABM is not an analog of object-oriented thing, object-oriented paradigms, however, it is a useful basis of ABM because agents can be treated as self-directed objects which have extraordinary action-selecting capacity. As a result, almost all ABM tools are object-oriented.

A simple ABM might have dozens or hundreds of agents, which can be used to learn how to build an ABM model and test the model. ABM is run in interpretative environment. Compiling or DLLs are not required. Ordinary environments include Java, etc.

A large scale ABM always has thousands or millions of agents. These ABM are usually operated in computer simulation environments. They support such characteristics as time dispatcher, communication mechanism, flexible interactive topology, some selective structures, components used to store and display the states of agents, etc.

There are already some standards for development of ABM platforms. For example, the Foundation for Intelligent Physical Agents' (FIPA 2005) architecture specifications, the Object Management Group Agent Platform Special Interest Group, Agent UML (OMG 2005), and the Knowledge-able Agent-oriented System architecture (KAoS) (Bradshaw, 1997).

So far some ABM platforms are available for users, including Start-logo (Resnick, 1994), Netlogo, Topographica (Bednar *et al.*, 2004), Swarm, TRANSIMS, CompuCell (Izaguirre *et al.*, 2004), NEURON (Hines and Carnevale, 1997), GENESIS (Wilson *et al.*, 1988), ABNNSim (Bonabeau, 2002), Evoland (Guzy *et al.*, 2008), etc.

StarLogo allows users to modeling using draggable GUI (Graphic User Interface). NetLogo is a branch of StarLogo, which allows users to modeling under a new Java GUI and redesigned modeling language (Fig. 1).

Swarm is the first general ABM tool based on Java. It is a software platform on the basis of agents, which provides some advantages for ecological modeling, including a set of standardized object bases, task scheduling base, detection base, and some structural characteristics as inheritance, message delivery, packaging, and hierarchical structures.

Echo was developed as a general model of the dynamics of adaptive systems. Echo is intended to embody mechanisms common to all adaptive systems. Holland gives an overview of attributes and mechanisms common to a range of adaptive systems, and builds a model whose mechanisms embody these tenets. The resulting model has similarities to extant modeling approaches, and also provides some innovative mechanisms which make it of particular value in the study of many-species dynamics. Echo is intended as a class of models whose members represent increasing complexity. As with the lattice-based models, Echo is an agent-based, spatially explicit model. Unlike a lattice-based model with one agent per node, populations of many agents interact within a site, and may migrate to other sites in a

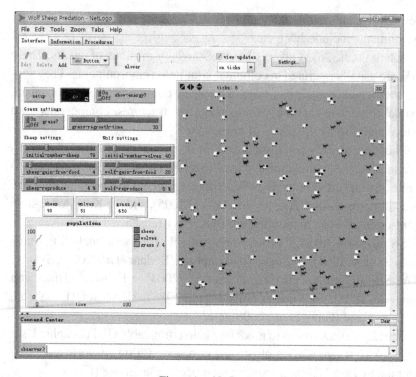

Figure 1.   NetLogo.

world of user-defined geometry. In addition to spatial and individual based
bookkeeping, Echo individuals have a genetic component. Agents have
their own haploid genomes, which are subject to heritable variation and
differential reproductive success. Thus, an analogy to evolution is possible,
along the lines of genetic algorithm simulations of evolutionary processes.
As they evolve from initial conditions, populations tend to be dominated
by genotypes with the greatest reproductive success. It is also possible to
create novel genotypes, or to model formation of novel species. This is a
stark contrast to extant many-species models, which either list all species
and determine their interactive attributes, or model evolution as a random
branching process. Echo genotypes are subject to an endogenous fitness
function. Endogenous fitness is made possible by resource-limited repro-
duction and genetically mediated behavior. Resources constitute the basic
currency of Echo, and are used to construct agent genotypes. An agent can
obtain resources from the environment or by interacting with other agents.

An agent may self-reproduce only when it has gathered sufficient resources to copy its genome. Genetically- mediated behavior determines whether two agents can interact.

TRANSIMS and CompuCell allow users to analyze the effects of different initial conditions on system's behaviors.

NEURON and GENESIS can exactly simulate the electrical and chemical nature of individual neurons. However, the size of the network that can be simulated is finite. GENESIS uses compartmental model to simulate biological neurons. Each section of neuron is simulated with path equation and the parameters in path equation include conductivity and capacity of input and output signals of cell. Similar to GENESIS, NEURON uses compartmental model to simulate biological neurons also. But the focus of NEURON is on modeling rather than compartment details. NEURON included higher GUI than GENESIS.

Topographica focuses on functions and large scale structures of neural networks.

Repast is a Java-based API (Application Programming Interface) which allows Java-based modeling. It is open-sourced and API components can be replaced or rewritten. Repast contains advanced displaying and analyzing tools and allows fast simulation (Fig. 2). Its fine time scheduler makes the time interval of discrete events reach double precision of Java. Repast Py is a platform-independent and visual model development system and allows fast ABM. It enables users to develop models based on GUI, and write behaviors of agents with Python script language. Repast J is a pure Java modeling environment. It includes a variety of features such as a fully concurrent discrete integration with geographical information systems using both Lagrangian and Eulerian representations for modeling agents on real maps, and adaptive behavioral tools such as neural networks and genetic algorithms. Repast J supports large scale ABM.

ABNNSim (Schoenharl, 2005; Fig. 3) was designed on the basis of Repast. It extends the electrical behaviors of neurons and includes the behavior for revising physical connections of a neuron to other neurons. Topological structure of network may thus change with time.

Evoland is used to simulate the changes of the environment under various policies in order to determine the regions that suffer from policy changes. The complete source codes, inputs guide and other details

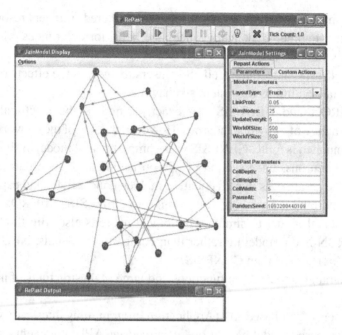

Figure 2.   Repast (Macal and North, 2005).

of Evoland 3.5 can be found at: http://evoland.bioe.orst.edu/. In the Evoland, an internal GIS manages various variables. GIS supports multiple languages, multiple selective desktop, and visualization pattern. Some attributes of Evoland are stored in the database. During the simulation the attributes of the database change with the change of external environment. Agents in Evoland are defined by a series of parameters. These parameters are used to evaluate health and economic value of the ecosystem studied. The attributes and parameters of spatial database are governed by a specific mechanism which is also used in cell automata. Different from cell automata, however, Evoland uses multi-dimensional space rather than a set of cells. In Evoland, the view is the unit of analysis. Each view consists of various operations. For each view, periodic cycling controls individual cycling and other non-individual patterns. Some of the patterns are called automatic programs. They can run without the participation of individuals. These automatic programs include the population growth and vegetation succession resulted from the changes of LULC. Another set of patterns

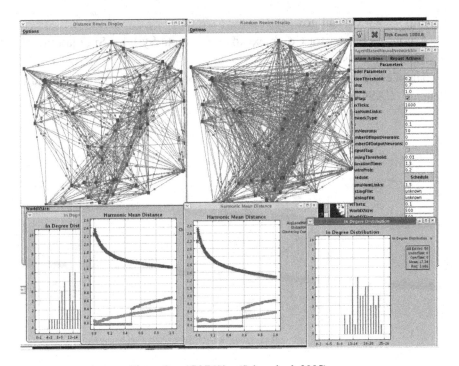

Figure 3.   ABNNSim (Schoenharl, 2005).

are landscape assessment models. These models calculate land's economic value and health condition of the ecosystem.

2.4.   *Resources for agent-based modeling*

The following are some of ABM resources and websites:

http://www.hcs.ucla.edu/arrowhead.htm

Focuses on social theory and modeling of human complex system.

http://www.ieeeswarm.org

The IEEE Swarm Intelligence Symposium.

http://www.santafe.edu

The Santa Fe Institute.

http://www.cscs.umich.edu

The Center for the Study of Complex Systems (CSCS)
at the University of Michigan.

http://www.cas.anl.gov
Argonne's Center for Complex Adaptive Agent Systems Simulation.
http://www.casos.cs.cmu.edu/
The Center for Computational Analysis of Social and Organizational Systems (CASOS) at Carnegie Mellon.

CHAPTER 12

Cell Automata Modeling of Pest Percolation

1. Introduction

Habitat heterogeneity refers to the diversity of soil, climate, vegetation, landscape and land use, etc. Habitat heterogeneity forms an important mechanism to retard pest infestation and formation of devastating pest population. Habitat heterogeneity helps to form stable food webs and ecosystems, and to form different types of barriers to pest infestation (Zhang and Schoenly, 1999). The study of pest infestation in heterogeneous habitat is important to sustainable management of pests.

Cell automata are a kind of agent-based modeling (individual based modeling; Griebeler, 2011). They have been widely used in emergent modeling and are proved to be a powerful tool (Qi and Zhang, 2002; Zhang, 2012). In the present study, we developed cell automata according to some basic rules of pest infestation. Various factors influencing pest infestation were studied using the cell automata and some general conclusions were drawn. The Java algorithm of the cell automata was also provided in the present study.

2. Cell Automata

Suppose the study region is divided into $c \times f$ grids with equal area, a_{ij}, where $i = 1, 2, \ldots, c$; $j = 1, 2, \ldots, f$, then pest infestation will occur in this region. Given the sensitivity level of each grid to pest infestation, $q_k(q_k \in R, k = 1, 2, \ldots, m)$, the corresponding probability of being infested $p_k(p_k \in [0, 1], k = 1, 2, \ldots, m)$, the initial distribution of pest infestation (that is, whether or not the pest occurs in a grid before the simulation ($t = 0$). The initial distribution is expressed as:

$$v_{ij} = 1, \quad \text{if grid } a_{ij} \text{ has been infested};$$
$$v_{ij} = 0, \quad \text{if grid } a_{ij} \text{ is not yet infested}.$$
$$i = 1, 2, \ldots, c; \quad j = 1, 2, \ldots, f,$$

and

$$v_{0i} = v_{f+1\,i} = 0, \quad i = 1, 2, \ldots, c + 1;$$
$$v_{i0} = v_{i\,c+1} = 0, \quad i = 1, 2, \ldots, f + 1.$$

Meanwhile, the sensitivity distribution is also known:

$$u_{ij} = q_k, \quad \text{if the sensitivity of grid } a_{ij} \text{ is } q_k (k = 1, 2, \ldots, m).$$
$$i = 1, 2, \ldots, c; \quad j = 1, 2, \ldots, f.$$

Thus the probability distribution of infestation is:

$$p_{ij} = p_k, \quad \text{if } u_{ij} = q_k (k = 1, 2, \ldots, m).$$
$$i = 1, 2, \ldots, c; \quad j = 1, 2, \ldots, f.$$

Suppose each infestation occurs at unit time, and the pest infects those grids with which the present grid has public edges. For any of the uninfected grid x, the pest will infect grid x with infestation probability p_i, from the present infected grid. For sensitivity k, calculate the number of grids available for the infestation, N_k:

$$N_k = \sum_i \sum_j b_{ij}$$

where,

$$b_{ij} = 1, \quad \text{if } (v_{ij} = 0) \wedge (p_{ij} = p_k) \wedge ((v_{i-1j} = 1) \vee (v_{i+1j} = 1)$$
$$\vee (v_{ij-1} = 1) \vee (v_{ij+1} = 1)) \text{ is true};$$
$$b_{ij} = 0, \quad \text{if } (v_{ij} = 0) \wedge (p_{ij} = p_k) \wedge ((v_{i-1j} = 1) \vee (v_{i+1j} = 1)$$
$$\vee (v_{ij-1} = 1) \vee (v_{ij+1} = 1)) \text{ is false}.$$

The number of actually infested grids with sensitivity k is:

$$n_k = \text{INT}((100 p_k N_k + 0.5)/100), \quad k = 1, 2, \ldots, m.$$

As a result, n_k actually infected grids can be generated among N_k grids.

On the other hand, the infested grid will restore to uninfected one at the restoration rate r (proportion of grids that can restore from infected state during unit time):

$$
\begin{aligned}
v_{ij}, & \quad \text{if } 1 - r + R \geq 1; \\
v_{ij} = 0, & \quad \text{if } 1 - r + R < 1. \\
& \quad i = 1, 2, \ldots, c; \quad j = 1, 2, \ldots, f.
\end{aligned}
$$

where R is a random value, $1 \geq R \geq 0$.

For each unit time, repeat the above iterative process, until the scheduled terminal time is achieved.

The main Java codes, PercoModel, are listed as:

```
/*c,f: numbers of grids along x-axis and y-axis; tm: duration
time of percolation; r: restoration probability; n: number of
sensitivity levels; nn1[]:sensitivity levels; nn2[]: probabili-
ties to be infected for specific sensitivity levels; u[][]:
sensitivity distribution of cells; v[][]: pest distribution*/
public class PercoModel {
public static void main(String[] args){
if (args.length!=6) System.out.println("You must input the
names of tables in the database. For example, you may type the
following in the command window: java PercoModel percomodel1
percomodel2 percomodel3 15 20 0.35, where percomodel1, per-
comodel2 and percomodel3 are the names of tables, 15 is the number
of column grids, 20 is the duration time of simulation, and 0.35 is
the restoration probability of cells. Sensitivity values of
grids are stored in table percomodel1. Values of percolation
sources are stored in the table percomodel2. Sensitivity levels
and corresponding probabilities to be infected are stored in the
table percomodel3.");
String tablename1 = args[0];
String tablename2 = args[1];
String tablename3 = args[2];
int c = Integer.valueOf(args[3]).intValue();
int tm = Integer.valueOf(args[4]).intValue();
```

```
double r = Double.valueOf(args[5]).doubleValue();
readDatabase readdata1=new readDatabase("dataBase",
tablename1,c);
readDatabase readdata2=new readDatabase("dataBase",
tablename2,c);
readDatabase readdata3=new readDatabase("dataBase",
tablename3,2);
int f = readdata1.m;
int n = readdata3.m;
int i, j, v[][];
double nn1[],nn2[],u[][];
u = new double[f+1][c+1];
v = new int[f+2][c+2];
nn1 = new double[n+1];
nn2 = new double[n+1];
for (i=1;i<=n;i++) {
nn1[i] = (Double.valueOf(readdata3.data[i][1])).
doubleValue();
nn2[i] = (Double.valueOf(readdata3.data[i][2])).
doubleValue(); }
for (i=1;i<=f;i++)
for (j=1;j<=c;j++) {
u[i][j] = (Double.valueOf(readdata1.data[i][j])).
doubleValue();
v[i][j] = (Integer.valueOf(readdata1.data[i][j])).
intValue(); }
perco(f, c, n, tm, r, nn1, nn2, u, v); }

public static void perco(int f, int c, int n, int tm,
double r, double nn1[], double nn2[], double u[][],
int v[][]) {
int i, j, k, nt;
int gr[][][],nn[],tt[],w[],cols[],nn3[];
double p[][],vv[];
vv = new double[tm+1];
nn = new int[n+1];
w = new int[f*c+1];
cols = new int[f*c+1];
p = new double[f+1][c+1];
gr = new int[tm+1][f+1][c+1];
tt = new int[tm+1];
nn3 = new int[n+1];
for (i = 1;i<=f;i++)
for (j = 1;j<=c;j++) gr[0][i][j] = v[i][j];
```

```
nt = 0;
graph(f, c, nt, v);
int c1 = 0;
for(i=1;i<=f;i++)
for(j=1;j<=c;j++)
if (v[i][j]!=0) c1++;
vv[0] = (double)c1/(f*c);
System.out.print("Percent grids infected:
"+String.valueOf((int)(vv[0]*10000)/100.00)+"%\n\n");
for(i=1;i<=f;i++)
for(j=1;j<=c;j++) {
double div = 1 - r+Math.random();
if (div>=1) gr[nt][i][j]=v[i][j];
else gr[nt][i][j]=0; }
for(i=1;i<=f;i++)
for(j=1;j<=c;j++)
v[i][j] = gr[nt][i][j];
tt[0] = 0;
for(i=1;i<=f;i++)
for(j=1;j<=c;j++)
for(k=1;k<=n;k++)
if (Math.abs(u[i][j]-nn1[k])<1.0e-08) p[i][j]=nn2[k];
for(i=1;i<=c+1;i++) {
v[0][i] = 0;
v[f+1][i] = 0; }
for(i=1;i<=f+1;i++) {
v[i][0] = 0;
v[i][c+1] = 0; }
for(nt=1;nt<=tm;nt++) {
for(k=1;k<=n;k++) {
nn3[k] = 0;
for(i=1;i<=f;i++)
for(j=1;j<=c;j++)
if ((v[i][j]==0) & (Math.abs(p[i][j]-nn2[k])<1.0e-08) &
((v[i - 1][j]==1) | (v[i+1][j]==1) | (v[i][j - 1]==1) |
(v[i][j+1]==1)))
nn3[k]++;
nn[k] = (int)((nn3[k]*nn2[k]*100+0.5)/100.00); }
for(k=1;k<=n;k++) {
int cc = 0;
for(i=1;i<=nn3[k];i++) w[i]=i;
while ((f!=0) | (c!=0)) {
int cs = (int)((nn3[k] - cc)*Math.random()+1);
cols[cc+1] = w[cs];
```

```
if (cs<nn3[k] - cc)
for(j=cs+1;j<=nn3[k] - cc;j++)
w[j - 1]=w[j];
cc++;
if (cc>=nn[k]) break; }
int pp = 0;
for(i=1;i<=f;i++)
for(j=1;j<=c;j++)
if ((v[i][j]==0) & (Math.abs(p[i][j]-nn2[k])<1.0e-08) &
((v[i-1][j]==1) | (v[i+1][j]==1) | (v[i][j-1]==1) |
(v[i][j+1]==1))) {
pp++;
for(int ii=1;ii<=nn[k];ii++)
if (cols[ii]==pp) v[i][j]=1; } }
tt[nt] = nt;
graph(f, c, nt, v);
c1 = 0;
for(i=1;i<=f;i++)
for(j=1;j<=c;j++)
if (v[i][j]!=0) c1++;
vv[nt] = (double)c1/(f*c);
System.out.print("Percent grids infected:
"+String.valueOf((int)(vv[nt]*10000)/100.00)+"%\n\n");
for(i=1;i<=f;i++)
for(j=1;j<=c;j++) {
double div = 1 - r+Math.random();
if (div>=1) gr[nt][i][j] = v[i][j];
else gr[nt][i][j] = 0; }
for(i=1;i<=f;i++)
for(j=1;j<=c;j++)
v[i][j] = gr[nt][i][j]; }
new GraphicsFrame(new PercoGraphics(0,f,c,tm,vv,tt,gr)).
resize(710,560); }

public static void graph(int f, int c, int nt, int v[][]) {
String ss = "";
System.out.print("------------------Time
"+String.valueOf(nt)+"----------------------\n");
for(int i=1;i<=f;i++) {
for(int j=1;j<=c;j++)
ss+=String.valueOf(v[i][j])+" ";
ss+="\n"; }
System.out.print(ss);
System.out.print("--------------------------------------"+"\n"); }
}
```

In the sensitivity file, the rows and columns are grids along y-axis and x-axis respectively. Each value in the cell is the sensitivity value of the grid:

```
0 2 1 3 0 0 3 2 0 2 1 1 0 0 1
3 0 1 2 0 3 0 2 1 1 3 0 0 2 0
0 0 0 3 1 0 1 0 3 0 0 3 1 2 0
0 0 1 0 0 0 2 0 1 1 0 0 2 3 0
0 2 1 3 0 1 0 3 2 1 1 2 3 2 2
3 0 1 2 0 3 0 4 0 0 0 4 0 3 0
0 0 3 1 0 0 1 2 0 2 2 2 1 0 0
0 0 1 3 0 0 1 2 0 2 2 0 0 2 1
2 1 1 2 0 2 0 3 2 1 3 0 0 2 0
0 0 0 3 1 0 1 0 0 0 0 3 0 2 0
0 0 1 0 0 0 2 0 1 1 0 0 0 3 0
0 2 1 3 0 0 0 3 0 0 1 2 0 2 0
3 0 1 2 0 3 0 0 0 0 0 2 0 3 0
0 0 3 1 0 0 1 2 0 0 2 2 1 0 2
0 0 1 2 0 0 1 2 0 2 3 0 0 2 1
```

In an initial infestation file, the rows and columns are grids along y-axis and x-axis respectively. Each cell has a value either 1 or 0. If the cell is the infected cell the value is 1, or else the value is 0:

```
0 1 1 0 0 0 1 0 0 1 0 1 0 0 1
1 0 1 0 0 1 0 1 0 1 0 0 0 1 0
0 0 0 0 1 0 1 0 1 0 0 1 1 1 0
0 0 1 0 0 0 1 0 1 1 0 0 1 0 0
0 0 1 0 0 1 0 1 1 1 1 0 0 0 1
0 0 1 0 0 1 0 0 0 0 0 0 0 0 0
0 0 0 1 0 0 1 0 0 0 0 0 1 0 0
0 0 1 0 0 0 1 0 0 0 0 0 0 1 0
0 1 1 1 0 0 0 1 0 1 0 0 0 0 0
0 0 0 0 1 0 1 0 0 0 0 1 0 1 0
0 0 1 0 0 0 0 0 1 1 0 0 0 1 0
0 0 1 0 0 0 0 1 0 0 1 0 0 0 0
1 0 1 0 0 1 0 0 0 0 0 0 0 0 0
0 0 0 1 0 0 1 0 0 0 0 0 1 0 0
0 0 0 0 0 0 1 0 0 0 0 0 0 0 1
```

In Infestation intensity file, each row has two values, the sensitivity of the grid and the corresponding probability to be infected.

The result output and graphic output of PercoModel are indicated in Fig. 1.

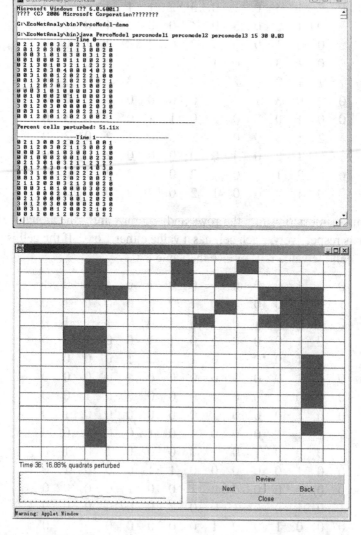

Figure 1.   Result output (upper) and graphic output (lower) of PercoModel.

3. Results

3.1. *Influence of restoration capacity on infestation process*

Suppose there is a region with 15 grids along both x-axis and y-axis. The terminal time for simulation is 40; there are four sensitivity levels ($m = 4$), i.e., $q_1 = 0, q_2 = 1, q_3 = 2, q_4 = 3$, and the corresponding infestation probability is $p_1 = 0, p_2 = 0.2, p_3 = 0.3, p_4 = 0.5$, respectively.

Set the restoration rate $r = 0, 0.05, 0.1, 0.15, 0.2, 0.25$, and 0.3, and simulate pest infestation process using the Java algorithm. The results are shown in Fig. 2.

Obviously, if all grids are not restorable, the proportion of infected grids will increase monotonically with time, and the upper limit of the proportion tends to be a certain value. One reason for the limited upper proportion is that the sensitivity of some grids is 0 (and infestation probability is 0) and the pest is not able to infest these grids. As the restoration rate increases, the proportion of infected grids decreases gradually. Once restoration rate reaches a certain level, the infected grids will disappear after a period of time. As can be seen from Fig. 2, the cell automata can describe the various types of infestation processes, such as the monotonous increase and decline, fluctuation, and periodic oscillation of infestation area. For example, the periodic oscillation in Fig. 2 may occur for $0.10 \geq r \geq 0.05$. For restoration rate $r = 0.075$, the periodic oscillation occurs (Fig. 3). As often met in practical situations, such periodic oscillation is not very strict. It has a certain time range, and is not so regular.

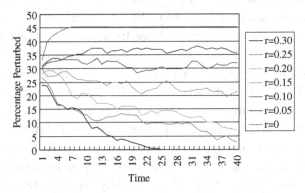

Figure 2.   Pest infestation process for restoration rate $r = 0, 0.05, 0.1, 0.15, 0.2, 0.25, 0.3$.

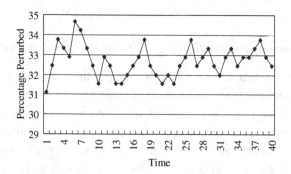

Figure 3.   Periodic oscillation of pest infestation for $r = 0.075$. The proportion of infected grids oscillates around 32.7% line.

The above results indicate that, restoration capability influences or even determines the pest infestation dynamics. Enhancing region's restoration capability is one of the most constructive measures to recover the region after being infected by the pest. There are many such examples. For example, the compensation capacity of rice and cotton has been reported to help these crops recover from pest injury.

Figure 4(A)–(C) show the distributions of pest infestation without restoration mechanism (other parameters unchanged).

3.2. *Influence of grid sensitivity on infestation process*

Sensitivity corresponds to infestation probability. A high level of sensitivity denotes a larger infestation probability, and vice versa. In the above example, there are four levels of sensitivity. We set restoration rate $r = 0$, and for the sensitivity level 0, 1, 2, and 3, we fix seven groups of infestation probabilities; for heterogeneous habitat, $(0, 0.1, 0.2, 0.3)$, $(0, 0.2, 0.3, 0.5)$, $(0, 0.3, 0.4, 0.6)$, $(0, 0.4, 0.5, 0.7)$, $(0, 0.5, 0.6, 0.8)$, and for homogeneous habitat, $(0.2, 0.2, 0.2, 0.2)$ and $(0.1, 0.1, 0.1, 0.1)$, and analyze the influence of grid sensitivity on infestation process. The results are shown in Fig. 5.

According to Fig. 5, infestation processes are much different between heterogeneous habitat and homogeneous habitat. Once a region in homogeneous habitat has a certain level of sensitivity, the pest would infect entire region in a period of time. If the grid sensitivity was high, then the upper limit of infestation area was high, the pest infected quickly and the time

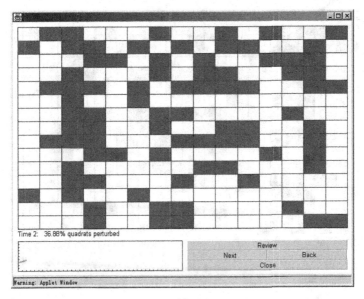

(a)

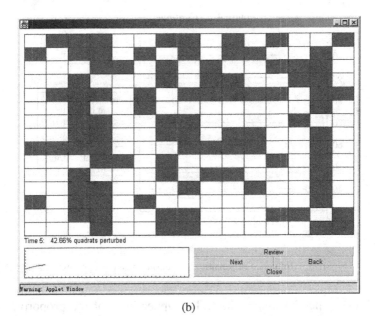

(b)

Figure 4. Distributions of pest infestation without restoration mechanism. (a) $t = 2$; (b) $t = 5$; (c) $t = 10$.

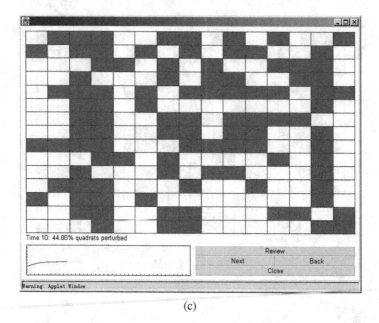

(c)

Figure 4. (*Continued*)

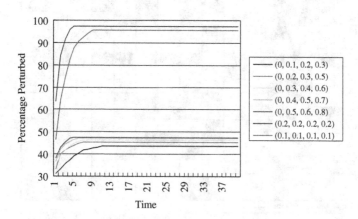

Figure 5. Influence of sensitivity level on infestation process.

to reach the upper limit was short. The upper limit of the proportion is about 43.5% to 47.6% in heterogeneous habitat. In heterogeneous habitat (or homogeneous habitat) the sensitivity of grids influence infestation process, but is not decisive.

3.3. *Influence of border barriers on infestation process*

Border barriers refer to the continuous grids that the pest cannot infect. The infestation probability (and sensitivity) is zero for these grids.

We fix four sets of infestation probabilities, (0, 0.2, 0.3, 0.5), (0, 0.3, 0.4, 0.6), (0.1, 0.2, 0.3, 0.5), and (0.1, 0.3, 0.4, 0.6), in which the latter two sets represent non-border barriers within the region (but it does not mean that the habitat is homogeneous because all grids have different levels of sensitivity), and restoration rate $r = 0$, or 0.1. The first two sets have border barriers and restoration rate $r = 0$. The influence of border barriers on infestation process is shown in Fig. 6.

Simulation results showed that with or without border barriers have a major impact on the infestation process. The presence of border barriers can effectively prevent pest infestation (Zhang and Schoenly, 1999). Border barriers and restoration capacity have decisive impacts on pest infestation. Border barriers have been widely used to prevent and control plant diseases and insect pests, for example, crop intercropping, multi-species layout, etc.

3.4. *Influence of layout of border barriers on infestation process*

Layout of border barriers refers to the location, shape, and size of border barriers in the field. In agricultural ecosystems, the layout of border barriers

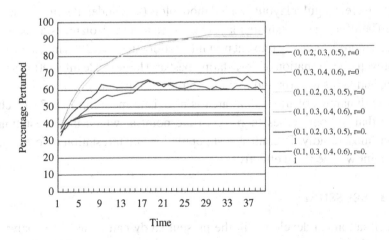

Figure 6.   Influence of border barriers on infestation process.

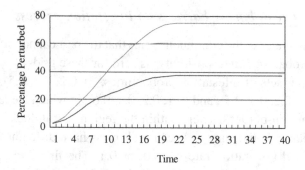

Figure 7.　Influence of layout of border barriers on infestation process. Upper curve: parallel layout; Lower curve: rectangular layout.

is usually regular. Thus we designed two layouts, i.e., the layout with parallel lines and the layout with rectangular shape. In parallel layout, a column of barrier grids (sensitivity $= 0$) was set per 4–5 columns of sensible grids (Sensitivity levels 1–3 are evenly distributed in all grids). In rectangular layout, the 8th row, 6th and 11th columns of grids (in total 15 rows and columns of grids, i.e., 225 grids in the region) are barrier grids and the others are sensible grids. Set restoration rate $= 0$, and the infestation probability for sensitivity 0–3 is 0 (border barrier grid), 0.3, 0.4 and 0.6. The results are shown in Fig. 7.

Clearly, different layout of border barriers will greatly affect pest infestation process and its outcome. Plots formed by parallel layout are fewer (four), and rectangular layout yielded more plots (six) under the same conditions. Therefore, the rectangular layout resulted in more isolated islands and thus can more effectively prevent pest infestation. In nature, this mechanism can prevent the formation of devastating pest and large-scale infestation and injury, and helps to form and maintain biological diversity.

The dynamics of infestation area is a logistic model in Fig. 7, which means that in some cases there is a time point at which the infestation spreads most rapidly. To control pest population and infestation before this, time point will be more effective.

4.　Discussion

The cell automata developed in the present study can be used to analyze effects of habitat diversity, i.e., mosaic pattern and plant distribution on pest infestation and population outcome; to determine the influence of pest infestation and injury on crop production, and to optimize the habitat pattern and plant distribution to reduce pest infestation and injury.

The resolution of grid partition should depend on the understanding of sensitivity levels and infestation probabilities of grids in the region studied, and the size of the region. A finer partition of region, i.e., more grids, will lead to high simulation accuracy. Methods of generating random numbers are different on different computers, so different computer systems will yield a slight different simulation results. This error can be reduced with the fining of grid partition.

More variables or parameters are expected in present cell automata so that this model can be used in more specific situations.

ABM Frame for Biological Community Succession and Assembly

One central question in food web theory is how structural and static patterns emerge from population-dynamics of interacting species (Cohen *et al.*, 1993). One possible way to explore this is to perform community assembly experiments. Most literature on assembly dynamics of multispecies ecosystems deals with competitive communities (Case, 1990; Morton *et al.*, 1996) or randomly wired ecosystems without trophic structure (May, 1973; Pimm, 1991). Assembly models of multispecies ecosystems with trophic structure have been less developed, starting from the early works (Pimm and Lawton, 1978; Pimm, 1979, 1980; Lockwood *et al.*, 1997; Bastolla *et al.*, 2001).

1. Ecosystem and Community

There are various levels in an ecosystem. Biologically, there are molecular ecology, individual ecology, population ecology, community ecology, ecosystem ecology, landscape ecology, etc. Biological communities are theme of community ecology.

Field observation of biological communities is always limited by the longer time. Understanding the mechanism of community composition and succession, has been the focus of ecologists (Hraber and Milne, 1997). Classical explanation of ecological processes assumed that communities were equilibrium systems that rarely experienced disturbances. Recovery from disturbances was expected to proceed in an orderly and linear way toward a stable state of a uniquely adapted species assemblage, i.e., a climax community. Species diversity, productivity, and stability were assumed to increase with time to a maximum at climax (Margalef, 1963; Savage and Askenazi, 1999). The intermediate disturbance hypothesis has been proposed which suggests that species diversity is highest, not at a stable endpoint but rather at an intermediate level of disturbances (frequency and intensity) (Margalef, 1963). Ecologists know that biological communities are basically non-equilibrium systems. A community is a self-organizing system generated from multiple species invasions, selection, adaptation, and optimization. External disturbances, for example, species invasion, might produce unpredictable and significant influence. Up till now, some processes or mechanisms governing community succession and assembly have been confirmed (Hraber and Milne, 1997):

(1) **Niche partition/Competitive exclusion.** One of two species that utilizes similar resources would be replaced by another due to their competitive interaction. This results from adaptive evolution that selectively utilizes resources, or the competitive exclusion between species.

(2) **Multiple attractors.** A community can have multiple distinct steady distributions or alternative steady states. They represent different species assemblages occurring at possible similar conditions. History of community succession determines which steady state will occur.

(3) **Spatial range.** Biological communities possess four orderliness: Number of species, number of individuals per species, space occupied by each species, and space occupied by each individual. Spatial heterogeneity, like resource aggregation or resource gradient, may reduce competition or predation effect by providing local refugee or fine adaptive mechanism. Environmental variation is indispensable to species richness, which results in different succession rules. However, spatial heterogeneity of species and individuals is always ignored by people.

(4) **Open drive system.** Biological communities are self-adaptive systems. They can respond to continuous fluctuations of the environment and population.

Biological communities are typical complex systems. Ecosystems, including biological communities, coincide with the characteristics of agents. Hence, ABM can be used for modeling spatial-temporal dynamics of ecosystems and biological communities.

So far, there are only few studies on how to use ABM in ecology. Hraber and Milne (1997) developed an ABM model on the basis of a self-adaptive system, Echo. It can be used to simulate the dynamic process of species assemblage. In this model, there are behaviors like predation, competition, and mating between individuals of different species or the same species. Different species and individuals have different genotypes and hence possess different fecundity and survival capacity. Fecundity and survival capacity might enhance by learning process. Genotypes of the species or individuals with greater fecundity and survival capacity are more easily multiplied. Spatial heterogeneity, however, is not considered in the model.

Topping *et al.* (2003) proposed an ABM model, ALMaSS, which was used to simulate the growth and spread of multiple species in the heterogeneous environment. This model does not consider the self-adaptive learning process of individuals, and there are only a few of between-species interaction types. The ABM model of Savage and Askenazi (1999), Arborscapes, can be used to describe competition, growth and spread processes of multiple tree species. Each individual in the model possesses some biological attributes and behavioral rules, and some disturbances like logging are also considered. As for between-species interactions, however, this model includes competition only. And self-adaptive learning process of individuals is ignored in the model. The ABM model (Fig. 1), developed by Qi *et al.* (2004), is a spatial heterogeneous model (cell automata), which includes spread and recovery process and can be used to describe population spread. Various spread dynamics and patterns will occur with changes in the topological structure of space and the recovery capacity.

2. ABM Frame for Biological Community Succession and Assembly

Based on previous research (see references in this book), the methodology for ABM of biological community is proposed as follows. It consists of three parts; acquire dynamic data on community succession, agent-based

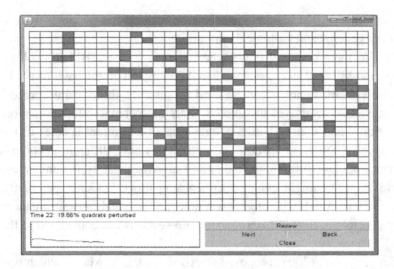

Figure 1.   An ABM model (cell automata) for population spread (Qi *et al.*, 2004).

modeling, and establishment of mechanism analysis. The main procedures are:

(1) Define various types of agents, and specify behaviors of agents.
(2) Identify relations between agents, and construct interaction types between agents.
(3) Select the platforms and environments for ABM, and set the strategies of ABM.
(4) Acquire necessary data for ABM. In the experiment and investigation of community dynamics, acquire spatial-temporal data of every species in the community. Use artificial recapture method to simulate species invasion and diffusion. Meanwhile, obtain some data from references and internet.
(5) Test the patterns of agents' behaviors and system's behaviors.
(6) Run ABM model, and analyze the output from the standpoint of linking the micro-scale behaviors of the agents to the macro-scale behaviors of the system.
(7) Analyze the mechanism of community succession and assembly using ABM model developed.

Some definitions and methods for ABM of community succession and assembly are described as follows.

2.1. *Agents and behaviors*

ABM can be based on existing modeling platforms, like Swarm and Echo. Because Swarm and Echo are inflexible, other platforms or methods can be additionally used, such as ALMaSS, Arborscapes, NetLogo, etc. Java is a pure object-oriented, distributed, robust, structure-neutral, platform-independent and dynamic programming language. UML is a general modeling language which supports not only object-oriented analysis and design, but also the entire process from requirement analysis, system design, to software implementation. UML provides strong support to software engineering. Both Java and UML have the common theoretical and ideological basis. We may perform systematic modeling and program using Java. Computation-extensive objects and methods can be realized as DLL (Dynamic Link Library). The available modeling environments include Windows and Linux operation systems. Modeling languages are UML, Java (JBuilder), Delphi (Borland Delphi), and C (Visual C++).

It is obvious that in a biological community, functional groups, species, individuals, etc., can be treated as agents at various levels.

Several types of agents can be defined as follows:

(1) **Species agents (invasive species agents and indigenous species agents).** Species agents include predator agents, parasitoid agents, neutral agents, herbivore agents (Fig. 2); or include agents that individuals of different species are just labeled with between-species coordination (positive or negative coordination, or neutral interaction (non coordination), magnitude of coordination). Between-species coordination can be derived from sampling data of species assemblages, expressed as partial correlation, coordination coefficient, etc.

(2) **Space agents.** They are represented by two-dimensional cells.

(3) **Functional agents.** They include interactive agents (user-model interactions), inductive agents (user induction of spatial dynamics, by such mechanisms as the change of distribution of plant resources (changing landscape structure)), and data collection and analysis agents. Of these agents, some agents may be designed as Java classes or DLLs using Java, Delphi, or C++.

In the model, the invasive species agent is treated as a common species agent in the community. Location and proportion of invasive species agent

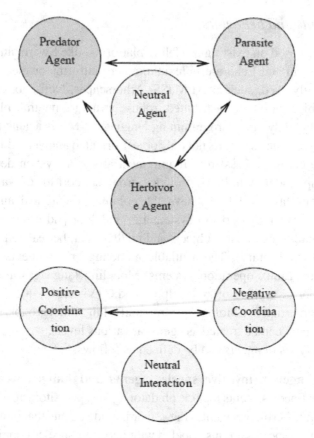

Figure 2.   Species agents and their interactions.

and frequency of invasive events are given fixed spatial or temporal probabilities.

The community is initialized with random specified number and location of individuals of every species, or initialized by investigated community data. The model proceeds on a month or annual or daily step basis.

The model includes plant resource input (herbivorous species agents must interact with plant resources). Herbivorous species agents are given an initial supply of each plant resource in its plant resource reservoir. A species agent may acquire resources from the environment or from the interactions with other species agents. Once enough resources are gathered, species agents can reproduce themselves.

Genetically-mediated behavior determines whether two species agents can interact. The species agent genome has two main regions, each with several attributes that code for a particular interaction. The *tag* region of the genome codes for attributes which are visible to other species agents. The *conditions* region represents attributes representing internal states, known only to the species agent. Matching a condition attribute against a tag attribute allows the interaction coded by those attributes to occur. Whether two species agents will interact is determined by comparing tag and condition attributes for that interaction. In the model, matching alleles in tag and condition attributes allows the interaction coded by those attributes. Tests for interactions are conducted sequentially: First for predation and competition, then for mutualism, and finally for mating.

Each species agent possesses some attributes: (1) a set of fixed, species-specific life history attributes, which include longevity, fecundity, age level, and others; (2) a time-relevant state (age, etc.).

Species agents are autonomous and have adaptive behavior, i.e., they adapt their behavior (growth, feeding, habitat selection, mate choice, etc.) according to their state and the environment, to seek higher fitness by using learning algorithms, e.g., ANNs (BP, Hebbian learning, etc.). Modeling adaptive system dynamics via individualistic mechanisms of adaptation is a fundamentally different approach than modeling with differential equations.

There are many behavioral models for agents. They include if-then rule and threshold model, artificial neural network and genetic algorithm rules, differential or difference equations, optimization rules, multivariable decision-making, etc.

The landscape dynamics depend on species agent interactions (competition, mutualism, predation, etc.) and dynamic landscape structure (plant resources distribution, landscape structure, etc.). System behavior depends on species agent interactions coupled with exogenous species agent. The spatial landscape of the model is a two-dimensional cell grid, in which each cell may be simultaneously occupied by many species (invasive species, indigenous species) (Fig. 3).

A space agent has major attributes including (Fig. 4): (1) plant resources availability; (2) number of individual agents of each species; (3) landscape structure and space available for species, etc. Landscape structure is driven

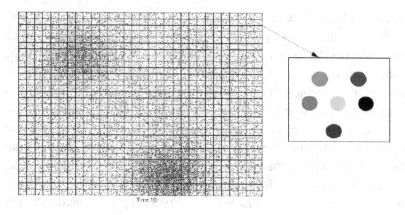

Figure 3.   Spatial gird cells.

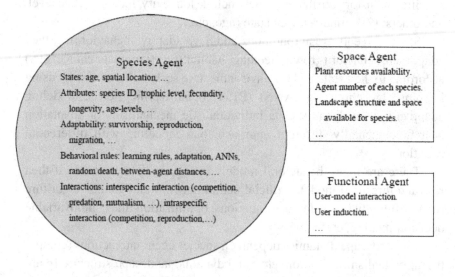

Figure 4.   Various agents and behavioral attributes.

by weather and other factors. Species agents migrate between grid cells according to the states of adjacent spatial cells.

In general, the state transition method and differential/ difference equations can be used to model the dynamics of species agents and (or) landscape structure.

2.2. Model objects

On ABM platform, agents are implemented as objects. The model is expected to include these objects: Species, GridCell, Individual, Invasion, LandStruc, etc. The simplest and most common object is the Individual, the record of the state of an individual of species, including its life history attributes. Few methods are implemented in the Individual class as the Species object is responsible for the actual state transitions that an individual may undergo. For example, an individual object is sent a method and then relays the message to its species object with a reference to itself as an argument. The GridCell object provides the space in which the simulation is taking place, a simple square grid where each cell represents the area required by some individuals. The GridCell has the following:

(1) A reference from every occupied cell in the grid to the individual currently residing at that location.
(2) A separate list of all the individuals currently residing within its boundaries.
(3) Current landscape structure and food availability. LandStruc generates dynamic landscape structure based on state transition method (or difference equation method), or just load it from GIS.

The Species object is the most complex participant in the model. Its role is to record all of the attributes of a species and to execute the actual simulation step on behalf of every individual belonging to the species. Attributes that differ across species include age levels, longevity, fecundity, trophic level, etc. The Invasion object is responsible for species invasion events in the model. Species invasions occur in space with uniform, random or aggregated distributions at pulse way. Other objects are designed to make user-model interactions, user induction, and data collection and analysis, etc.

2.3. Visualization tools

In addition to defining agents and a schedule of events, the model can provide visual tools for the observation of the model on a time step basis. Windows will graphically track the abundance and spatial distribution of each species including invasive species, etc. A probe feature allows the user to query any cell on species, age, and attributes of the individual at the site. These window functions can be crucial to evaluating model behavior. Windows allow modification of parameters including size and structure of the landscape, number of species, frequency of invasion, etc.

2.4. *Pattern analysis*

A pattern is anything above random variation and thus indicates some kind of internal organization. Pattern-oriented ABM starts with identifying a variety of observed patterns, at different scales and at both individual and system levels, that characterize the system's dynamics and mechanisms. These patterns, along with the problem being addressed and conceptual models of the system, provide the basis for designing and testing our ABM. To analyze spatial and temporal pattern dynamics generated by the model, the output can be linked to one of many pattern analysis packages. We may also write pattern analysis algorithms (spatial distribution patterns, topological structures, etc.) into the code. The model output may be linked to Fragstats (a comprehensive pattern analysis program). Its raster version is appropriate for the evaluation of the structure of cell-based models, where it generates metrics for area, patch density, size and variability, edge, shape, core area, diversity, contagion, interspersion, and nearest neighbor values. On the other hand, we may design pattern analysis algorithms (spatial distribution pattern, topological structure (shape, size, mosaic density, boundary, connectedness, etc.)).

Identifying the critical invasion strength for community may help develop ways to manage landscapes to maintain a particular ecological function, or to make comparisons between different communities. The critical invasion strength for community can be determined by the method of spatial phase transitions.

2.5. *Parameterization*

A major problem of ABM of real systems is parameterization. Parameters are acquired from community investigation or experiments or internet. Many parameters would be uncertain or even unknown. Consequently, model results are uncertain and predictions and insights from the model become questionable. Sensitivity analysis with limited available parameter values will provide a partial solution.

2.6. *Model application*

The completed model will be used to approach patterns and mechanisms of community succession and assembly.

Usually, the steps of the ABM have to be repeated several times as this will lead to new theories, additional patterns, or modification of the entire ABM.

❧ CHAPTER 14 ❧

Agent-based Modeling of Ecological Problems

1. Evolution Games

An understanding of evolution games is useful for agent-based modeling of biological evolution and community succession.

1.1. *Evolution theory and cooperation*

Evolutionary theory is based on competition for existence and survival of the fittest. According to evolutionary theory, evolution occurs at the level of individuals but not at the level of population. Hence it is natural to assume that individuals are selfish in the natural selection. However, natural phenomena and social processes have shown that cooperation, which is an adaptation to nature and environment, exists widely among individuals of same category, and even among individuals of distinct categories.

Some improvements on evolutionary theory include:

(1) Dawkins (1976) assumed that natural selection occurs at the level of genes.

(2) For a group of individuals with close genetic relationship, if the loss of a single individual helps to survive other individuals in the group, then this cooperation is a genetic adaptation to nature. Most cooperation behaviors of biological systems (but not human) occurs among individuals with close genetic relationship.

(3) Axelrod and Hamilton (1981) held that cooperation occurs at the level of individuals. He assumed that each individual pursues its own interest; there is not a central individual to force them cooperating. By using a game model, The Prisoner's Dilemma, it was found that cooperation, occurred in a small group of individuals exhibiting retribution behavior, might produce and evolve in a selfish world without central individuals. It was pointed out that cooperation will evolve if and only if it is based on retribution and the future effect should be so important that the retribution is always stable.

1.2.   *Parrondo's Paradox Game*

1.2.1.   *Parrondo's Paradox Game*

Parrondo's Paradox Game Theorem (Harmer and Abbott, 1999a,b): Given two games, if the losing probability of each game is greater than its winning probability, then a winning result will likely be obtained when the two games occur randomly or periodically.

The Parrondo's Paradox Game was based on the following statements:

(1) Biological individuals possess two properties, population property and nature property. Population property is relevant to competition and cooperation among individuals in the population. Nature property is relevant to the effect of its environment on itself.

(2) The model, The Prisoner's Dilemma, only simulates the game relationship among individuals (populations). It does not consider the role of the environment in the evolution of individuals or populations. Consequently, it was supposed that there are two game relationships in the evolution process of individuals, i.e., the game between individuals, and the game between individuals and the environment. The Parrondo's Paradox could be used to simulate the evolution of individuals and analyze the generation and development of between-individual cooperation.

1.2.2.   *An evolution model based on Parrondo's Paradox*

A model based on the Parrondo's Paradox Game was developed to describe biological evolution (Toral, 2001, 2002; Xie *et al.*, 2010; Fig. 1).

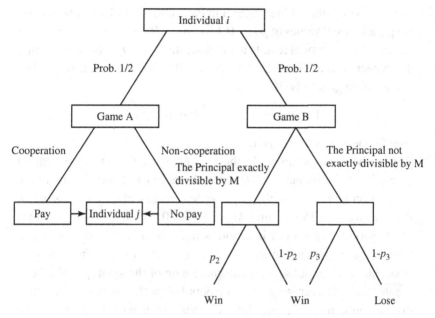

Figure 1.   A biological evolution game based on Toral interpretation of Parrondo's Paradox Game (Xie *et al.*, 2010).

There are two game relationships in the evolution of individuals:

(1) **Between-individual zero-sum game (Game A).** The gain of an individual is just the loss of another individual. This mechanism fully reflects the competition and selfishness of individuals in the evolution.

(2) **Negative game between individuals and the environment (Game B).**

Consider a population consisting of N individuals. The game is as follows (Xie *et al.*, 2010):

(1) Randomly select an individual i. For this individual, randomly select Game A or Game B with the same probability, 0.5. Game A means that if individual i selects to cooperate then pay a unit of its gain to randomly selected individual j; if individual i selects not to cooperate then it need not pay anything. In the view of population, Game A is a fair zero-sum game. It only changes the assignment pattern of gains in the population.

(2) Game B is the same as the Game B in Parrondo's Paradox Game theory. Set parametrical values of $p_2 = 0.1-\varepsilon$, $p_3 = 0.75-\varepsilon$, and $M = 3$, $\varepsilon = 0.005$. The theoretical results using discrete Markov chain showed that the expectation of Game B is $E(B) = -0.008695$. Thus, it is a losing game. More generally, if

$$(1 - p_2)(1 - p_3)^{M-1}/(p_2 p_3^{M-1}) > 1,$$

then Game B is a losing game.

(3) Repeat the game K times. In the computer simulation, set the initial gain of each individual as 200 units, the number of individuals in the population as $r = 200$, the parametrical values of Game B as $p_2 = 0.1 - \varepsilon$, $p_3 = 0.75 - \varepsilon$, and $M = 3$, $\varepsilon = 0.005$, and repeating times $K = 84000$. Population invasion is defined in such a way, suppose there exists a population that holds a specific strategy; there exists a small mutative population; if the game gain of the small population is greater than the average gain of the population, then the small mutative population can invade population, otherwise it will disappear in the evolution process.

Simulation results showed that the cooperation of all individuals helps to the existence and development of the population (positive gain; Fig. 2). The perfect non-cooperation of all individuals (all selfish individuals) will result in natural elimination of entire population (negative gain; Fig. 2).

The results showed that a small group of cooperating individuals (cooperating group) may effectively invade the non-cooperating population i.e., the average gain of a group of cooperating individuals is greater than that for the non-cooperating population. Furthermore, the larger the small group (more individuals), the greater is its average gain. When the number k of individuals in the small group is not less than 3.5% of entire population, its average gain is positive (>200 units), and thus the small group will survive and evolve in the population.

It was found that the conservatively cooperating population is stable and can avoid the invasion by a non-cooperating group (all individuals in the group are selfish). With increase in the size of the non-cooperating group, the average gain of the conservatively cooperating population tends to decline. Thus, in the view of the conservatively cooperating population, the non-cooperating group is unfavorable to both itself and the entire population.

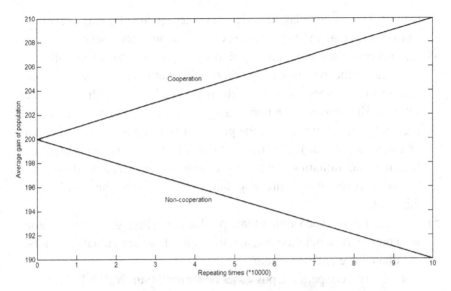

Figure 2.   Results of perfect cooperation and non-cooperation.

Open cooperating population is unstable and a small non-cooperating group can successfully invade the population. The average gain of the small non-cooperating group is significantly greater than that of entire population. So non-cooperation behavior is attractive to the individuals in an open cooperating population.

Hints and conclusions drawn from the above model include:

(1) Most research on biological competition and cooperation is based on The Prisoner's Dilemma model. Nevertheless, they did not consider the effect of natural selection on individuals' behavior. Moreover, the setting of game gains (i.e., gain from cooperation is greater than the gain from non-cooperation) led to a certain motivation to cooperation strategy in the process of long term interactions. Hence the conclusions on the basis of this model cannot reflect the fact that cooperation is an adaptation of individuals to nature.

(2) Parrondo's Paradox Game model represents two practical game relationships: between-individual zero-sum game (Game A), and negative game between individuals and nature (Game B).

(3) Species might evolve like a ratchet wheel. When a cooperation emerges occasionally, the environment might easily destroy the cooperation

pattern. The factors playing as ratchet wheel (Game B) may prevent it from destroying, and help species evolve to more complexity.

(4) The process for cooperative evolution may be described as follows. First, due to the existence of mutative individuals, a conservative cooperating group emerges in the selfish population (all individuals are selfish). The conservative cooperating group then achieves an advantage in the competition (i.e., the gain of this group is greater than that of entire population). Through natural selection, between-individual learning and imitation, and between-generation heredity, the cooperation spreads over entire population. Cooperation thus evolves in this way.

(5) Cooperation between individuals produces positive gain in the game with nature. A population consisting of all selfish individuals might be eliminated (i.e., negative gain).

(6) Conservative cooperation possesses better initial survival ability. A conservatively cooperating group might invade and survive in a selfish population. Conservative cooperation is stable, thus a selfish group is not able to invade a conservatively cooperating population.

(7) Provided some action is adopted to facilitate Game A (e.g., active cooperation between individuals, or cruel fighting between individuals) and thus creating a non-equilibrium among the individuals, then driving by ratchet wheel mechanism of Game B, entire population will achieve a positive gain and evolve to more complexity.

(8) Both between-individual cooperation in small groups and cruel competition among individuals jointly facilitate the evolution of population. This is why competition and cooperation can co-exist in the evolution. Both of them are adaptation to nature.

2. ABM of Some Ecological Problems

2.1. *IBMs of ecological and evolutionary processes*

DeAngelis and Mooij (2005) described the individual-based modeling (IBM) of ecological and evolutionary processes. In their research, five major types of individual-based models (IBMs) were proposed, i.e., spatial, ontogenetic, phenotypic, cognitive, and genetic. IBM can be divided into five aspects: (1) spatial diversity; (2) life cycles and ontogenesis; (3) community diversity (community adaptation, community behaviors); (4) differences in experiences and learning; (5) genetic and evolutionary diversity.

It was proposed that the state variables of an IBM can be visualized as a table in which the rows represent the individuals and the columns represent their traits. A typical IBM keeps track of between 100 and 10,000 organisms.

2.2. A modeling framework for investigating impact of landscape structure changes on population dynamics

Wiegand *et al.* (1999) constructed a general modeling framework that allows for a systematic investigation of the impact of changes in landscape structure on population dynamics (Fig. 3). The framework included a landscape generator with independent control over landscape composition

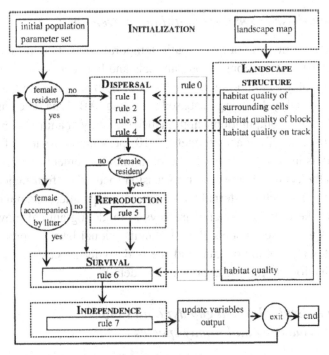

Figure 3.   Flow chart for the individual-based simulation model of European brown bear population (from Wiegand *et al.*, 1999).

and hysiognomy, an individual-based spatially explicit population model that simulates population dynamics within heterogeneous landscapes, and scale-dependent landscape indices that describe scale-dependent correlation between and within habitat types, (Wiegand *et al.*, 1999).

In their framework, landscape maps are represented by a grid of 50 by 50 cells and consist of good quality, poor quality, or uninhabitable matrix habitat cells. The population model was shaped in accordance to the biology of European brown bears, and demographic parameters were adjusted to yield a source-sink configuration. It was found that landscape indices were able to explain variations in variables of population dynamics caused by different landscape structure. When landscape structure changed, changes in these variables generally followed the corresponding change of an appropriate landscape index in a linear way. This framework incorporated source-sink dynamics and meta- population dynamics, and the population model can easily be modified for other species groups.

2.3.   *Agent-based modeling of animal movement*

Animal movement is a complex spatio-temporal phenomenon (Tang and Bennett, 2010). Interactions among animals and between animals and the environment play an important role in the development of the complex ecological and social systems. Agent-based modeling (ABM) has been increasingly applied as a computational approach to the study of animal movement across landscapes. Tang and Bennett (2010) presented a review of agent-based models in which the simulation of animal movement processes and patterns is the central theme. Their discussion focused on four major components: Internal states, external factors, motion capacities, and navigation capacities. The four components are keys in modeling animal movement behavior. Agent-based models allow for an individual-based approach that encapsulates these components, and the underlying processes that drive animal behavior can thus be approached in detail.

2.4.   *An individual-based gap dynamics model of herbaceous and woody species*

In his study on a grassland–shrubland transition zone, Peters (2002) presented a mixed life form individual plant-based gap dynamics model

(ECOTONE) to examine consequences of differences in recruitment, resource acquisition, and mortality to patterns in species dominance and composition under a variety of soils and climatic conditions. This model represents interactions among multiple potential dominant species that include congeneric species of one life form as well as herbaceous and woody life forms across multiple spatial scales. Similar to other gap models, ECO-TONE simulates the recruitment, growth, and mortality of individual plants on a small plot through time at an annual time step. According to Peters (2002), ECOTONE differs from other gap models in the degree of detail involved in determining successful recruitment by each species and in the simulation of below-ground resources. Individual plant root distributions and resource availability by depth are dynamic. Soil water content is simulated on a daily time step, and nitrogen is simulated monthly. Multiple spatial scales can be simulated using a grid of plots connected by seed dispersal. It was found that the individual-based modeling approach is capable of representing complex interactions among herbaceous and woody species as well as between congeneric species with different life history traits at a biome transition zone. This modeling approach is useful in improving understanding of key processes driving these vegetation dynamics as well as in predicting shifts in dominance as environmental conditions change in the future.

2.5. A swarming intelligence and agent-based modeling approach

Pérez and Dragićević (2010) used an intelligent agent-based model (ABM) to capture the complexity of the Mountain Pine Beetle (MPB) emergence, aggregation and attack behavior. Agent-based approach permits simulation of interactions that describe the ecological context in which insect populations spread. In their methods, intelligent reasoning was introduced by a swarm intelligence (SI) algorithm integrated with the ABM that describes indirect communication, collective behavior and self-organized aggregation of insects in a forest ecosystem. The model, ForestSimMPB, developed by them was calibrated by fine tuning two model parameters, and is implemented using recorded data. Simulation outputs provided predictions of spatial patterns in the forest landscape structure as a result of a MPB outbreak. The model can be used to improve methods for prevention and control of MPB disturbances.

2.6.　*An agent-based and analytical modeling to evaluate the effectiveness of greenbelts*

Brown *et al.* (2004) proposed an agent-based and analytical modeling to evaluate the effectiveness of greenbelts. Three agent-based models that include agents with the potential for heterogeneous preferences and a landscape with the potential for heterogeneous attributes were developed. Furthermore, they presented two different two-dimensional ABMs and conducted a series of experiments to supplement mathematical analysis. These include examining the effects of heterogeneous agent preferences, multiple landscape patterns, incomplete or imperfect information available to agents, and a positive aesthetic quality impact of the greenbelt on neighboring locations.

2.7.　*Agent-based modeling and evolution of eco-industrial systems*

Cao *et al.* (2009) applied agent-based modeling to eco-industrial systems for gaining new insights into their behavior. The factory in the eco-industrial systems was taken as an agent. The objects, attributes and behaviors were determined and some important interaction mechanisms between agents were designed. Moreover, the sustainability evolution was studied using emergy theory. A new concept, the Internal-flow emergy, was used to depict the evolution of an eco-industrial system. A hypothetical eco-industrial park utilizing natural gas and halite as the main raw material inputs was adopted as a case study to illustrate the effectiveness of the agent-based modeling.

2.8.　*Agent-based modeling of deforestation and reforestation*

Manson and Evans (2007) combined mixed-methods research with integrated agent-based modeling to understand land change and economic decision making in the United States and Mexico.

Their work demonstrated how sustainability science benefits from combining integrated agent-based modeling and mixed-methods research (which interleaves multiple approaches ranging from qualitative field research to quantitative laboratory experiments and interpretation of remotely sensed imagery). They have tested assumptions of utility-maximizing behavior in household-level landscape management in south-central Indiana. They also used evolutionary programming to represent bounded rationality in agriculturalist households in the southern Yucatán, Mexico. This approach was proved to capture realistic rule of thumb strategies while identifying social and environmental factors in a manner similar to econometric models. Their studies highlighted the role of computational models of decision making in land-change contexts.

2.9. An agent-based modeling for evaluating effect of variable fishing strategy on fisheries

Cabral *et al.* (2010) used an agent-based model to evaluate the response of a two-species fish community to fishing boat exploration strategies (boats following high-yield boats (Cartesian), boats fishing at random sites (stochast-random), and boats fishing at least exploited sites (stochast-pressure)).

They found that at low fishing pressure, the stochast-random mode yielded a high average catch per boat while sustaining fish biomass. At high fishing pressure, the Cartesian mode was more effective. For the Cartesian strategy, fish biomass exhibited four distinct behaviors with increasing number of boats. In order to break the pattern of localization (bandwagon effect), they introduced stochast-random intruders in a Cartesian-dominated fishery. Adding a single intruder changed the patchy-structured stock biomass pattern of a purely Cartesian fishery to a uniformly explored stock biomass pattern because of the additional spatial information provided by the intruder. The average catch per boat increased but at the expense of a disproportionate declined in equilibrium biomass (Cabral *et al.*, 2010).

2.10. An agent-based model for assessing human effect on wolf behavior

Musiani *et al.* (2010) have developed an agent-based model to investigate how varying levels of human presence could affect wolf behaviors, including highway crossings; use of areas in proximity to roads and trails; size of home ranges; activities, such as hunting, patrolling, resting, and feeding pups; and survival of individuals in Banff and Kootenay National Parks, Canada.

The model consisted of a wolf module as the primary component with five packs represented as cognitive agents, and grizzly bear, elk, and human modules that represent dynamic components of the environment (Musiani *et al.*, 2010). A set of environmental data layers was used to develop a friction model that serves as a base map representing the landscape over which wolves moved. A decision model was built to simulate the sequence of wolf activities. The model was implemented in a Java Programming Language using RePast, an agent-based modeling library. Wolf activities were simulated and calibrated with GPS data from wolf radiocollars deployed from 2002 to 2004.

Results showed that the simulated trajectories of wolf movements were correlated with the observed trajectories; other critical behaviors, such as time spent at the den and not traveling were also correlated. The simulations showed that wolf movements and behaviors were significantly affected by the intensity of human presence. The packs' home ranges shrank and wolves crossed highways less frequently with increased human presence. The modeling prototype developed in their study were considered to serve as a tool to test hypotheses about human effects on wolves and on other mammals, and guide decision-makers in designing management strategies that minimize impacts on wolves and on other species functionally related to wolves in the ecosystem (Musiani *et al.*, 2010).

References

Achlioptas D, D'Souza RM, Spencer J. Explosive percolation in random networks. *Science*, **323**: 1453–1455, 2009.

Albert R, Jeong H, Barabasi AL. Error and attack tolerance of complex networks. *Nature*, **406**: 378–382, 2000.

Amundsen PA, Lafferty KD, Knudsen R, *et al.* Food web topology and parasites in the pelagic zone of a subarctic lake. *Journal of Animal Ecology*, **78**: 563–572, 2009.

Arai HP, Mudry DR. Protozoan and metazoan parasites of fishes from the headwaters of the Parsnip and McGregor Rivers, British Columbia: a study of possible parasite transfaunations. *Canadian Journal of Fisheries and Aquatic Sciences*, **40**: 1676–1684, 1983.

Arthur JR, Margolis L, Arai HP. Parasites of fishes of Aishihik and Stevens Lakes, Yukon Territory, and potential consequences of their interlake transfer through a proposed water diversion for hydroelectrical purposes. *Journal of the Fisheries Research Board of Canada*, **33**: 2489–2499, 1976.

Axelrod R, Hamilton WD. The evolution of cooperation. *Science*, **211**: 379–403, 1981.

Barabasi AL. Scale-free networks: a decade and beyond. *Science*, **325**: 412–413, 2009.

Barabasi AL, Albert R. Emergence of scaling in random networks. *Science*, **286**(5439): 509, 1999.

Bascompte J. Networks in ecology. *Basic and Applied Ecology*, **8**: 485–490, 2007.

Bascompte J. Disentangling the web of life. *Science*, **325**: 416–419, 2009.

Bascompte J, Jordano P. Plant-animal mutualistic networks: the architecture of biodiversity. *Annual Review of Ecology, Evolution and Systematics*, **38**: 567–593, 2007.

Bastolla U, Fortuna MA, Pascual–Garcia A, *et al.* The architecture of mutualistic networks minimizes competition and increases biodiversity. *Nature*, **458**: 1018–1020, 2009.

Bastolla U, Lassig M, Manrubia SC, Valleriani A. Diversity patterns from ecological models at dynamical equilibrium. *Journal of Theoretical Biology*, **212**: 11–34, 2001.

Batagelj V, Mrvar A. Pajek 2.00: Reference manual. Ljubljana, 2010.

Beehler B. Frugivory and polygamy in birds of paradise. *The Auk*, **100**(1): 1–12, 1983.

Bednar JA, Choe Y, Paula JD, *et al.* Modeling cortical maps with topographica. *Neurocomputing*, **58**: 1129–1135, 2004.

Bellman RE. *Dynamic Programming*. Princeton University Press, Princeton, USA, 1957.

Berlow EL, Jennifer A, Dunne A, *et al.* Simple prediction of interaction strengths in complex food webs. *Proceedings of the National Academy of Sciences of U.S.A*, **106**: 187–191, 2009.

Berlow EL, Navarrete SA, Briggs CJ, *et al.* Quantifying variation in the strengths of species interactions. *Ecology*, **80**: 2206–2224, 1999.

Bluthgen N, Stork NE, Fiedler K. Bottom-up control and co-occurrence in complex communities: honeydew and nectar determine a rainforest ant mosaic. *Oikos*, **106**: 344–358, 2004.

Bohman T. Emergence of connectivity in networks. *Science*, **323**: 1438–1439, 2009.

Bollobás B. *Random Graphs*. Academic Press, USA, 1985.

Bollobas B. *Random Graphs*. Cambridge University Press, Cambridge, UK, 2001.

Bonabeau E. Agent-based modeling: methods and techniques for simulating human systems. *Proceedings of the National Academy of Sciences of U.S.A*, **99**(3): 7280–7287, 2001.

Borgatti S. *NetDraw* 2.099. Lexington, USA, 2011.

Bradshaw J. *An Introduction to Software Agents*. AAAI Press, CA, USA, 1997.

Brose U, Williams RI, Martinez ND. Allometric scaling enhances stability in complex food webs. *Ecology Letters*, **9**: 1228–1236, 2006.

Brown DG, Page SE, Riolo R, Rand W. Agent-based and analytical modeling to evaluate the effectiveness of greenbelts. *Environmental Modelling and Software*, **19**: 1097–1109, 2004.

Bruno JF, Stachowicz JJ, Bertness MD. Inclusion of facilitation into ecological theory. *Trends in Ecology and Evolution*, **18**(3): 119–125, 2003.

Butts CT. Revisiting the foundations of network analysis. *Science*, **325**: 414–416, 2009.

Cabral RB, Geronimo RC, Lim MT, Alino PM. Effect of variable fishing strategy on fisheries under changing effort and pressure: An agent-based model application. *Ecological Modelling*, **221**: 362–369, 2010.

Callaway RM. Positive interactions among plants (Interpreting botanical progress). *The Botanical Review*, **61**: 306–349, 1995.

Cancho RF, Sole RV. *Optimization in complex networks*. Santa Fe Institute, USA, 2001.

Cao K, Feng X, Wan H. Applying agent-based modeling to the evolution of eco-industrial systems. *Ecological Economics*, **68**: 2868–2876, 2009.

Case TJ. Invasion resistance arises in strongly interacting species-rich model competition communities. *Proceedings of the National Academy of Sciences of U.S.A.*, **87**: 9610–9614, 1990.

Casti J. *Would-Be Worlds: How Simulation Is Changing the World of Science*. Wiley, New York, USA, 1997.

Cattin MF, Bersier LF, Banasek-Richter C, Baltensperger R, Gabriel JP. Phylogenetic constraints and adaptation explain food-web structure. *Nature*, **427**: 835–839, 2004.

Chan SB, *et al. Graph Theory and Its Applications*. Science Press, Beijing, China, 1982.

Chan JX. *Lectures on Foundations of Algebraic Topology*. Higher Education Press, Beijing, China, 1987.

Chen X, Cohen JE. Global stability, local stability and permanence in model food webs. *Journal of Theoretical Biology*, **212**: 223–235, 2001.

Chen Y, Hu XC, Liu BB, *et al*. Modeling and simulation of economic system based on UML and Swarm. *Computer Engineering and Design*, **29**(15): 4040–4042, 4060, 2008.

Christensen V, Pauly D. ECOPATH II — a software for balancing steady-state ecosystem models and calculating network characteristics. *Ecological Modelling*, **61**(3–4): 169–185, 1992.

Christensen V, Walters CJ. Ecopath with Ecosim: methods, capabilities and limitations. *Ecological Modelling*, **172**: 109–139, 2004.

Cohen JE. *Food Webs and Niche Space*. Princeton University Press, Princeton, NJ, USA, 1978.

Cohen JE, *et al*. Improving food webs. *Ecology*, **74**: 252–258, 1993.

Cohen JE, Beaver RA, Cousins SH, *et al*. Improving food webs. *Ecology*, **74**: 252–258, 1993.

Cohen JE, Briand F. Trophic links of community food webs. *Proceedings of the National Academy of Sciences of U.S.A.*, **81**: 4105–4109, 1984.

Cohen JE, Briand F, Newman CM. *Community Food Webs: Data and Theory*. Springer, Berlin, Germany, 1990.

Cohen JE, Newman CM. A stochastic theory of comunity food webs. I. Models and aggregated data. *Proceedings of the Royal Society of London Series B*, **224**: 421–448, 1985.

Coleman BD, Mares MA, Willig MR, *et al*. Randomness, area, and species richness. *Ecology*, **63**: 1121–1133, 1982.

Dawkins R. *The Selfish Gene*. Oxford University Press, UK, 1976.

DeAngelis DJ, Mooij WM. Individual-based modeling of ecological and evolutionary processes. *Annual Review of Ecology, Evolution and Systimatics*, **36**:147–168, 2005.

Dickerson JE, Robinson JV. Microcosms as islands: a test of the MacArthur–Wilson equilibrium theory. *Ecology*, **66**: 966–980, 1985.

Dijkstra EW. A note on two problems in connection with graphs. *Numerche Math*, **1**: 269–271, 1959.

Dormann CF. How to be a specialist? Quantifying specialisation in pollination networks. *Network Biology*, **1**(1): 1–20, 2011.

Dunn WR Jr, Chan SP. An algorithm for testing the planarity of a graph. *IEEE Trans on Circuit Theory*, **15**(2): 166–168, 1968.

Dunne JA, Williams RJ, Martinez ND. Food–web structure and network theory: the role of connectance and size. *Proceedings of the National Academy of Sciences of U.S.A.*, **99**(20): 12917–12922, 2002.

Dunne JA, Williams RJ, Martinez ND. Network structure and biodiversity loss in food webs: robustness increases with connectance. *Ecology Letters*, **5**: 558–567, 2002.

Dunne JA, Williams R, Martinez N. Conservation of species interaction networks. *Ecology Letters*, **5**: 558, 2002.

Editorial. Ecological network theory. *Ecological Modelling*, **208**: 1–2, 2007.

Erdos P, Renyi A. On random graphs. *Publicationes Mathematicae Debrecen*, **6**: 290–297, 1959.

Fath BD. Network analysis applied to large-scale cyber-ecosystems. *Ecological Modelling*, **171**: 329–337, 2004.

Fath BD, Patten BC. Quantifying resource homogenization using network flow analysis. *Ecological Modelling*, **123**: 193–205, 1999a.

Fath BD, Patten BC. Review of the foundations of network environ analysis. *Ecosystems*, **2**: 167–179, 1999b.

Fath BD, Scharler UM, Ulanowiczd RE and Hannone B. Ecological network analysis: network construction. *Ecological Modeling*, **208**: 49–55, 2007.

Faust K. Very local structure in social networks. *Sociological Methodology*, **37**: 209–256, 2007.

Fautin DG, Allen GR. Field guides to anemone fishes and their host sea anemones. Western Australian Museum, Australia Read phonetically Dictionary, 1997.

Fecit. *Analysis and Design of Neural Networks in MATLAB* 6.5. Electronics Industry Press, Beijing, China, 2003.

Finn JT. Measures of ecosystem structure and function erived from analysis of flows. *Journal of Theoretical Biology*, **56**: 363–380, 1976.

Finn JT. Cycling index: a general definition for cycling incompartment models. *Environmental Chemistry and Cycling Processes*, (Eds.) Adriano DC, Brisbin IL. Vol. 45, *DOE Proceedings*, Conf. 760429. National Technical Information Service, Springfield, VA, 1978, pp. 138–165.

Finn JT. Flow analysis of models of the Hubbard Brook ecosystem. *Ecology*, **61**: 562–571, 1980.

FIPA (Foundation for Intelligent Physical Agents). FIPA Home Page, 2005. http://www.fipa.org/

Floyd RW. Algorithm 97: shortest path. *Comm ACM*, **5**: 345, 1962.

Ford LR Jr, Fulkerson DR. Maximal flow through a network. *Canadian Journal of Mathematics*, **8**: 399–404, 1956.

Ford LR Jr, Fulkerson DR. A simple algorithm for finding maximal network flow and application to the Hitchcock problem. *Canadian Journal of Mathematics*, **9**: 210–218, 1957.

Gallai T. On directed paths and circuits. *Theory of Graphs*, Frdos P, Katona G (Eds.). Academic Press, New York, USA, 115–118, 1968.

Gardner M. The fantastic combinations of John Conway's new solitaire game Life. *Scientific American*, **223**: 120–123, 1970.

Griebeler EM. 2011. Are individual based models a suitable approach to estimate population vulnerability? A case study. *Computational Ecology and Software*, **1**(1): 14–24, 2005.

Grimma V, Berger U, Bastiansen F. A standard protocol for describing individual-based and agent-based models. *Ecological Modelling*, **198**: 115–126, 2006.

Gross JL. Yellen J. *Graph Theory and Its Applications* (2nd edn.), Chapman & Hall/CRC, USA, 2005.

Guzy MR, Smith CL, Bolte JP, Hulse DW, Gregory SV. Policy research using agent-based modeling to assess future impacts of urban expansion into farmlands and forests. *Ecology and Society*, **13**(1): 37, 2008.

Halnes G, Fath BD, Liljenstrom H. The modified niche model: including detritus in simple structural food web models. *Ecological Modelling*, **208**: 9–16, 2007.

Hannon B. The structure of ecosystems. *Journal of Theoretical Biology*, **41**: 535–546, 1973.

Hannon B. Ecosystem flow analysis. *Ecological Theory for Biological Oceanography*, Ulanowicz R, Platt T (Eds.). 97–118.

Hannon B. Ecosystem control theory. *Journal of Theoretical Biology*, **121**: 417–437, 1986.

Hannon B. Empirical cyclic stabilization of an oyster reef ecosystem. *Journal of Theoretical Biology*, **149**: 507–519, 1991.

Hannon B. Ecological pricing and economic efficiency. *Ecological Economy*, **36**: 19–30, 2001.

Hannon B, Costanza R, Herendeen R. Measures of energy cost and value in ecosystems. *Journal of Environmental Economics and Management*, **13**: 391–401, 1986.

Hannon B, Costanza R, Ulanowicz R. A general accounting framework for ecological systems: a functional taxonomy for connectivist ecology. *Theoretical Population Biology*, **40**(1): 78–104, 1991.

Hannon B, Joiris C. A seasonal analysis of the Southern North Sea ecosystem. *Ecology*, **70**: 1916–1934, 1989.

Harmer GE, Abbott D. Parrondo's paradox. *Statistical Science*, **14**(2): 206–213, 1999a.

Harmer GP, Abbott D. Losing strategies can win by Parrondo's paradox. *Nature*, **402**(6764): 864–870, 1999b.

Havens K. Scale and structure in natural food webs. *Science*, **257**: 1107–1109, 1992.

Higashi M, Patten BC. Further aspects of the analysis of indirect effects in ecosystems. *Ecological Modelling*, **31**: 69–77, 1986.

Higashi M, Patten BC. Dominance of indirect causality in ecosystems. *American Naturalist*, **133**: 288–302, 1989.

Hines ML, Carnevale NT. The neuron simulation environment. *Neural Computation*, **9**: 1179–1209, 1997.

Holland JH. *Adaptation in Natural and Artificial Systems: An Introductory Analysis with Applications to Biology, Control and Artificial Intelligence*. University of Michigan Press, Ann Arbor, MI, USA, 1975.

Hopcroft J, Tarjan RE. Efficient planarity testing. *ACM J*, **12**(4): 549–568, 1974.

Hraber PT, Milne BT. Community assembly in a model ecosystem. *Ecological Modelling*, **103**: 267–285, 1997.

Huspeni TC, Lafferty KD. Using larval trematodes that parasitize snails to evaluate a salt-marsh restoration project. *Ecological Applications*, **14**: 795–804, 2004.

Huxham M, Raffaelli D, Pike A. Parasites and food–web patterns. *Journal of Animal Ecology*, **64**: 168–176, 1995.

Ibrahim SS, Eldeeb MAR, Rady MAH, *et al.* The role of protein interaction domains in the human cancer network. *Network Biology*, 1(1): 59–71, 2011.

Izaguirre JA, Chaturvedi R, Huang C, *et al.* CompuCell, a multimodel framework for simulation of morphogenesis. *Bioinformatics*, 20(7): 1129–1137, 2004.

Jaynes ET. Information theory and statistical mechanics. *Physics Review*, 106: 620–630, 1957.

Jennings NR. On agent-based software engineering. *Artificial Intelligence*, 117: 277–296, 2000.

Jiang QF. *Practical Decision-making Analysis.* Guizgou People's Press, Guiyang, China, 1988.

Joern A. Feeding patterns in grasshoppers (Orthoptera: Acrididae): factors influencing diet specialization. *Oecologia*, 38: 325–347, 1979.

Jordán F. Trophic fields. *Community Ecology*, 2: 181–185, 2001.

Jordán F, Liu W, Davis AJ. Topological keystone species: Measures of positional importance in food webs. *Oikos*, 112: 535–546, 2006.

Jorgensen SE, Fath B. Examination of ecological networks. *Ecological Modelling*, 196: 283–288, 2006.

Kondoh M. Building trophic modules into a persistent food web. *Proceedings of the National Academy of Sciences of U.S.A.*, 105: 16631–16635, 2008.

Krebs CJ. *Ecological Methodology* (2nd ed.). Benjamin Cummings, Menlo Park, California, USA, 1999.

Kuang WP, Zhang WJ. Some effects of parasitism on food web structure: a topological analysis. *Network Biology*, 1(3–4): 171–185, 2011.

Kuris AM. Guild structure of larval trematodes in molluscan hosts: prevalence, dominance and significance of competition. *Parasite Communities: Patterns and Processes*, Esch GW, Bush AO, Aho JM (Eds.). Chapman and Hall, London, UK, 69–100, 1990.

Lafferty KD, Dobson AP, Kuris AM. Parasites dominate food web links. *Proceedings of the National Academy of Sciences of the USA*, 103(30): 11211–11216, 2006a.

Lafferty KD, Hechinger RF, Shaw JC, *et al.* Food webs and parasites in a salt marsh ecosystem. *Disease Ecology: Community Structure and Pathogen Dynamics*, Collinge S, Ray C (Eds.). Oxford University Press, Oxford, UK, 119–134, 2006b.

Lafferty KD, Sammond DT, Kuris AM. Analyses of larval trematode communities. *Ecology*, 75: 2275–2285, 1994.

Laska MS, Wooton JT. Theoretical concepts and empirical approaches to measuring interaction strength. *Ecology*, 79: 461–476, 1998.

Latham LG. Network flow analysis algorithms. *Ecological Modelling*, 192: 586–600, 2006.

Latham LG, Scully EP. Quantifying constraint to assess development in ecological networks. *Ecological Modelling*, 154: 25–44, 2002.

Law R, Blackford JC. Self-assembling food webs. A global view-point of coexistence of species in Lotka–Volterra communities. *Ecology*, 73: 567–578, 1992.

Leong TS, Holmes JC. Communities of metazoan parasites in open water fishes of Cold Lake, Alberta. *Journal of Fish Biology*, 18: 693–713, 1981.

Li HL, Chen H, Jin SY. Research of modeling method for Agent-based complex systems distributed simulation. *Computer Engineering and Applications*, **43**(8): 209–213, 2007.

Li, *et al. Operational Research*. Tsinghua University Press, Beijing, China, 1982.

Li Y, Ma SF. Modeling of agent-based simulation system. *Journal of Systems Engineering*, **21**(3): 225–231, 2006.

Libralato S, Christensen V, Pauly D. A method for identifying keystone species in food web models. *Ecological Modeling*, **195**: 153–171, 2006.

Lin JK. *Foundations of Topology*. Science Press, Beijing, China, 1998.

Lockwood JL, Powell RD, Nott MP, Pimm SL. Assembling ecological communities in space and time. *Oikos*, **80**: 549–553, 1997.

Lu KC, Lu HM. *Graph Theory and Its Applications* (2nd edn.). Tsinghua University Press, Beijing, China, 1995.

Luo HF. An optimization analysis on cellular metabolic network model. MS Dissertation, Sun Yat-sen University, China, 2007.

MacArthur R. Fluctuation of animal populations and a measure of community stability. *Ecology*, **36**(3): 533–536, 1955.

Macal CM, North MJ. Tutorial on agent-based modeling and simulation. *Proceedings of the 2005 Winter Simulation Conference*, Kuhl ME, Steiger NM, Armstrong FB, Joines JA (Eds.). 2005.

Manly BFJ. *Randomization, Bootstrap and Monte Carlo Methods in Biology* (2nd Edn.). Chapman & Hall, London, UK, 1997.

Margalef R, Manson SM, Evans T. Agent-based modeling of deforestation in southern Yucatan, Mexico, and reforestation in the Midwest United States. *Proceedings of the National Academy of Sciences of U.S.A.*, **104**(52): 20678–20683, 2007.

Martinez ND. Artifacts or attributes? Effects of resolution on the Little Rock Lake food web. *Ecological Monographs*, **61**: 367–392, 1991.

Martinez ND. Constant connectance in community food webs. *American Naturalist*, **139**: 1208–1218, 1992.

Martinez ND. Scale-dependent constraints on food web structure. *American Naturalist*, **144**: 935–953, 1994.

Martinez ND, Hawkins BA, Dawah HA, Feifarek B. Effects on sampling effort on characterization of food-web structure. *Ecology*, **80**: 1044–1055, 1999.

Martinez-Antonio A. *Escherichia coli* transcriptional regulatory network. *Network Biology*, **1**(1): 21–33, 2011.

Mason SJ. Feedback theory-some properties of signal flow graphs. *Proceeding of the Institute of Radio Engineers*, **41**: 1144–1156, 1953.

May RM. Will a large complex system be stable. *Nature*, **238**: 413, 1972.

May RM. *Stability and Complexity in Model Ecosystems*. Princeton University Press, USA, 1973.

May RM. The structure of food webs. *Nature*, **301**: 566–568, 1983.

McCann KS. The diversity-stability debate. *Nature*, **405**: 228–233, 2000.

McCann KS, Hastings A, Huxel GR. Weak trophic interactions and the balance of nature. *Nature*, **395**: 794–798, 1998.

Mellouli S, Mineau G, *et al*. Laying the foundations for an agent modelling methodology for fault tolerant multi-agent systems. *Fourth International Workshop Engineering Societies in the Agents World*. Imperial College London, UK, 2003.

Memmott J, Waser NM, Price MV. Tolerance of pollination networks to species extinctions. *Proceedings of the Royal Society of London Series B*, **271**: 2605–2611, 2004.

Milo R, *et al*. Network motifs: simple building blocks of complex networks. *Science*, **298**: 824–827, 2002.

Minty GJ. A simple algorithm for listing all the trees of a graph. *IEEE Trans on Circuit Theory*, CT,**12**(1): 120, 1965.

Montoya JM, Pimm SL, Sole RV. Ecological networks and their fragility. *Nature*, **442**: 259–264, 2006.

Montoya JM, Sole RV. Small world patters in food webs. *Journal of Theoretical Biology*, **214**: 405–412, 2002.

Montoya JM, Sole RV. Topological properties of food webs: from real data to community assembly models. *Oikos*, **102**: 614–622, 2003.

Morin PJ, Lawler SP. Effects of food chain length and omnivory on population dynamics in experimental food webs. *Food Webs: Integration of Patterns and Dynamics*, Polis GA, Winemiller KO (Eds.). Chapman and Hall, 218–230, 1996.

Morris JT, Christian RR, Ulanowicz RE. Analysis of size and complexity of randomly constructed food webs by information theoretic metrics. *Aquatic Food Webs: An Ecosystem Approach*, Belgrano A, Scharler UM, Dunne J, Ulanowicz RE (Eds.). Oxford University Press, Oxford, UK, 73–85, 2005.

Morton D, Law R, Pimm SL, *et al*. On models for assembling ecological communities. *Oikos*, **75**: 493–499, 1996.

Morton D, Law R, Pimm SL, Drake JA. On models for assembling ecological communities. *Oikos*, **75**: 493–499, 1996.

Montoya JM, Pimm SL, Sole RV. Ecological networks and their fragility. *Nature*, **442**: 259–264, 2006.

Murtaugh PA, Kollath JP. Variation of trophic fractions and connectance in food webs. *Ecology*, **78**: 1382–1387, 1997.

Musiani M, Anwar Sk M, McDermid GJ, Hebblewhite M, Marceau DJ. How humans shape wolf behavior in Banff and Kootenay National Parks, Canada. *Ecological Modelling*, **221**: 2374–2387, 2010.

National Center for Ecological Analysis and Synthesis. Interaction Web Database [DB/QL] (2011). www.nceas.ucsb.edu/interactionweb/html/datasets.html.

Navia AF, Cortés E, Mejía-Falla PA. Topological analysis of the ecological importance of elasmobranch fishes: A food web study on the Gulf of Tortugas, Colombia. *Ecological Modelling*, **221**: 2918–2926, 2010.

NetLogo. (2004) http://http://ccl.northwestern.edu/netlogo.

NetLogo. (2005) http://ccl.northwestern.edu/netlogo.

Newman N, Barabasi A, Warrs DJ. *The Structure and Dynamics of Networks*. Princeton University Press, Princeton, NJ, USA, 2006.

Norton M. *Modern Control Engineering*. Pergamon Press, UK, 1972.

Ollerton J, Johnson SD, Cranmer L, *et al*. The pollination ecology of an assemblage of grassland asclepiads in South Africa. *Annals of Botany*, **92**: 807–834, 2003.

Ollerton J, McCollin D, Fautin DG, *et al*. Finding NEMO: nestedness engendered by mutualistic organisation in anemonefish and their hosts. *Proceedings of the Royal Society of London Series B*, **274**: 591–598, 2007.

Paine RT. Food-web analysis through field measurements of per capita interaction strengths. *Nature*, **355**: 73–75, 1992.

Park YS, Sovan L, Scardi M, Versonshot PFM, Jorgensen SE. Patterning exergy of benthic macroinvertebrate communities using self-organising maps. *Ecological Modelling*, **190**: 105–113, 2006.

Pascual M, Dunne JA. *Ecological Networks: Linking Structure to Dynamics in food Webs*. Oxford Univ. Press, Oxford, UK, 2006.

Paton K. An algorithm for finding a fundamental set of cycles of a graph. *Communications of the ACM*, **12**(9): 514–518, 1969.

Patten BC, Bosserman RW, Finn JT, Cale WG. Propagation of cause in ecosystems. In *Systems Analysis and Simulation in Ecology*, Patten BC (Ed.). Academic Press, New York, Vol. 4, pp. 457–579, 1976.

Patten BC. Systems approach to the concept of environment. *Ohio Journal of Science*, **78**: 206–222, 1978.

Patten BC. Environs: the superniches of ecosystems. *American Zoologist*, **21**: 845–852, 1981.

Patten BC. Environs: relativistic elementary particles or ecology. *American Naturalist*, **119**: 179–219, 1982.

Patten BC. Energy cycling in the ecosystem. *Ecological Modelling*, **28**: 1–71, 1985.

Pauly D, Christensen V, Walters C. Ecopath, Ecosim, and Ecospace as tools for evaluating ecosystem impacts on marine ecosystems. *ICES Journal of Marine Science*, **57**: 697–706, 2000.

Petchey OL, Beckerman AP, Riede JO, *et al*. Size, foraging, and food web structure. *Proceedings of the National Academy of Sciences of U.S.A.*, **105**(11): 4191–4196, 2008.

Petchey OL, Eklof A, Borrvall C, Ebenman B. Trophically unique species are vulnerable to cascading extinction. *American Naturalist*, **171**: 568–579, 2008.

Peters PC. Plant species dominance at a grassland–shrubland ecotone: an individual-based gap dynamics model of herbaceous and woody species. *Ecological Modelling*, **152**: 5–32, 2002.

Pérez L, Dragićević S. ForestSimMPB: A swarming intelligence and agent-based modeling approach for mountain pine beetle outbreaks. *Ecological Informatics*, **6**(1): 62–72, 2011.

Pimm SL. Complexity and stability: another look at MacArthur's original hypothesis. *Oikos*, **35**: 139–149, 1979.

Pimm SL. Food web design and the effect of species deletion. *Oikos*, **35**: 139–149, 1980.

Pimm SL. *Food Webs*. Chapman & Hall, London, UK, 1982.

Pimm SL. *The Balance of Nature*. University of Chicago Press, Chicago, USA, 1991.

Pimm SL. *The Balance of Nature?: Ecological Issues in the Conservation of Species and Communities*. University of Chicago Press, USA, 1991.

Pimm SL, Lawton JH. On feeding on more than one trophic level. *Nature*, **275**: 542–544, 1978.

Pimm SL, Lawton JH. Are food webs divided into compartments? *Journal of Animal Ecology*, **49**: 879–898, 1980.

Pimm SL, Lawton JH, Cohen JE. Food web patterns and their consequences. *Nature*, **350**: 669–674, 1991.

Pinnegar JK, Blanchard JL, Mackinson S, *et al.* Aggregation and removal of weak-links in food-web models: system stability and recovery from disturbance. *Ecological Modelling*, **184**: 229–248, 2005.

Polis GA. Complex trophic interactions in deserts: an empirical critique of food web theory. *American Naturalist*, **138**: 123–155, 1991.

Polovina JJ. Model of a coral-reef ecosystem. 1. The Ecopath model and its application to French Frigate Shoals. *Coral Reefs*, **3**: 1–11, 1984.

Poulin B, Wright SJ, Lefebvre G, *et al.* Interspecific synchrony and asynchrony in the fruiting phenologies of congeneric bird-dispersed plants in Panama. *Journal of Tropical Ecology*, **15**: 213–227, 1999.

Qi YH. The web computational software for one-dimensional ordered cluster analysis of sequential information. *Information Science*, **23**(Suppl.): 99–101, 2005.

Qi YH, Zhang WJ. A percolation model for pest perturbation in diverse habitat and net work computing software. *Modern Computer*, **133**: 16–19, 2002.

Qi YH, Zhang WJ, Zhang ZG. Stochastic percolation model and analysis for pest infestation in the heterogeneous habitat. *Computer Applications Research*, **24**(Suppl.): 299–301, 2004.

Raffaelli DG, Hall SJ. Assessing the relative importance of trophic links in food webs. *Food Webs: Integration of Patterns and Dynamics*, Polis GA and Winemiller KO (Eds.). Chapman and Hall, 185–191, 1996.

Repast (2004) http://repast.sourceforge.net/.

Resnick M. *Turtles, Termites and Traffic Jams*. MIT Press, USA, 1994.

Rezende EL, Lavabre JE, Guimaraes PR, *et al.* Nonrandom coextinctions in phylogenetically structured mutualistic networks. *Nature*, **448**: 925–928, 2007.

Richardson TL, Jackson GA, Burd AB. Planktonic food web dynamics in two contrasting regions of Florida Bay, US. *Bulletin of Marine Science*, **73**: 569–591, 2003.

Roy B. Nomber chromatique et plus longs chemins d'un graphe. *Rev Francaise Automat Informat Recberche Operationelle Ser Rouge*, **1**: 127–132, 1967.

Rutledge RW, Basorre BL, Mulholland RJ. Ecological stability: an information theory view-point. *Journal of Theoretical Biology*, **57**: 355–371, 1976.

Savage M, Askenazi M. Arborscapes: a Swarm-based Multi-agent Ecological Disturbance Model (1999). http://citeseerx.ist.psu.edu/viewdoc/summary?doi=10.1.1.50.1188.

Schoenharl TW. An agent based modeling approach for the exploration of self-organizing neural networks. MS Thesis, University of Notre Dame, USA, 2005.

Schoenly K, Beaver RA, Heumier TA. On the trophic relations of insects: a food web approach. *American Naturalist*, **137**: 597–638, 1991.

Schoenly KG, Zhang WJ. IRRI Biodiversity Software Series. I. LUMP, LINK, AND JOIN: Utility programs for biodiversity research. *IRRI Technical Bulletin No. 1*. Manila (Philippines): International Rice Research Institute, Manila, Philippines, 1999a.

Schoenly KG, Zhang WJ. IRRI Biodiversity Software Series. V. RARE, SPPDISS, and SPPANK: programs for detecting between-sample difference in community structure. *IRRI Technical Bulletin No.5*. International Rice Research Institute, Manila, Philippines, 1999b.

Shannon CE. *The theory and design of linear differential equation. Machines, OSRD Rept. 411, Sec. D-2 (Fire Control) of the U.S. National Defence Research Committee*, USA, 1942.

Slegers MF, Stroosnijder L. Beyond the desertification narrative: a framework for agricultural drought in semi-arid east Africa. *Ambio*, **37**(5): 372–380, 2008.

Small E. Insect pollinators of the Mer Bleue peat bog of Ottawa. *Canadian Field Naturalist*, **90**: 22–28, 1976.

Sole RV, Montoya JM. Complexity and fragility in ecological networks. *Proceedings of the Royal Society of London Series B*, **268**: 2039–2045, 2001.

Solow AR. A simple test for change in community structure. *Journal of Animal Ecology*, **62**: 191–193, 1993.

Spanier EH. *Algebraic Topology*. Springer-Verlag, New York, USA, 1966.

Sporns O. *Networks of the Brain*. MIT Press, Cambridge, USA, 2010.

Sporns O, Tononi G, Edelman GM. Theoretical neuroanatomy: relating anatomical and functional connectivity in graphs and cortical connection matrices. *Cerebral Cortex*, **10**(2): 127–141, 2000.

Sprecht DF. Probabilistic neural networks for classification, mapping and associative memory. *IEEE ICNN*, San Diego, USA, 1988.

Sprecht DF. Probabilistic neural networks. *Neural Networks*, **3**(1): 109–118, 1990.

Sprules WG, Bowerman JE. Omnivory and food chain length in zooplankton food webs. *Ecology*, **69**: 418–426, 1988.

StartLogo. (2004) http://www.media.mit.edu/starlogo.

Stouffer DB, Camacho J, Jiang W, Amaral LAN. Evidence for the existence of a robust pattern of prey selection in food webs. *Proceedings of the Royal Society of London Series*, **274**: 1931–1940, 2007.

Sugihara G, Schoenly K, Trombla A. Scale invariance in food web properties. *Science*, **245**: 48–52, 1989.

SWARM. (2004) http://wiki.swarm.org/.

Tacutu R, *et al.* Immunoregulatory network and cancer-associated genes: molecular links and relevance to aging. *Network Biology*, **1**(2): 112–120, 2011.

Tang WW, Bennett DA. Agent-based modeling of animal movement: a review. *Geography Compass*, **4**(7): 682–700, 2010.

Tarjan RE. Depth-first search and linear graph algorithms. *SIAM Journal on Computing*, **1**(2): 146–160, 1972.

Thompson JN. *The Geographic Mosaic of Coevolution*. University of Chicago Press, Chicago, USA, 2005.

Thompson RM, Mouritsen KN, Poulin R. Importance of parasites and their life cycle characteristics in determining the structure of a large marine food web. *Journal of Animal Ecology*, **74**: 77–85, 2005.

Topping CJ, Hansen TS, Jensen TS, *et al.* ALMaSS, an agent-based model for animals in temperate European landscapes. *Ecological Modeling*, **167**: 65–82, 2003.

Toral R. Cooperative Parrondo's games. *Fluctuation and Noise Letters*, **1**: 7–12, 2001.

Toral R. Capital redistribution brings wealth by Parrondo's paradox. *Fluctuation and Noise Letters*, **2**: 305–311, 2002.

Ulanowicz RE. An hypothesis on the development of natural communities. *Journal of Theoretical Biology*, **85**: 223–245, 1980.

Ulanowicz RE. Identifying the structure of cycling in ecosystems. *Mathematical Biosciences*, **65**: 219–237, 1983.

Ulanowicz RE. *Growth and Development, Ecosystems Phenomenology*. Springer, New York, USA, 1986.

Ulanowicz RE. Ecosystem trophic foundations: lindeman exonerata. *Complex Ecology: The Part-Whole Relation in Ecosystems*, Patten BC, Jørgensen SE (Eds.), Prentice-Hall, Englewood Cliffs, New Jersey, USA, 549–560, 1995.

Ulanowicz RE. Ecology, the ascendent perspective. *Complexity in Ecological Systems Series*, Allen TFH, Roberts DW (Eds.), Columbia University Press, New York, USA, 1997.

Ulanowicz RE. A synopsis of quantitative methods for ecological network analysis. *Computational Biology and Chemistry*, **28**(5–6): 321–339, 2004.

Ulanowicz RE, Kemp WM. Toward a canonical trophic aggregation. *American Naturalist*, **114**: 871–883, 1979.

Ulanowicz RE, Norden JS. Symmetrical overhead in flow networks. *International Journal of Systems Science*, **21**: 429–437, 1990.

Ulanowicz RE, Puccia CJ. Mixed trophic impacts in ecosystems. *Coenoses*, **5**(1): 7–16, 1990.

Ulanowicz RE, Kay JJ. A pacakage for the analysis of ecosystem flow networks. *Environmental Software*, **6**: 131–142, 1991.

Vezina AF, Pahlow M. Reconstruction of ecosystem flows using inverse methods: how well do they work? *Journal of Marine Systems*, **40–41**: 55–77, 2003.

Vezina AF, Platt TC. Food web dynamics in the ocean. *I. Best Estimates of Flow Networks Using Inverse Methods. Marine Ecology Progress Series*, **42**: pp. 269–287.

Walters CJ, Christensen V, Pauly D. Structuring dynamic models of exploited ecosystems from trophic mass-balance assessments. *Reviews in Fish Biology and Fisheries*, **7**: 139–172, 1997.

Walters CJ, Kitchell JF, Christensen V, Pauly D. Representing density dependent consequences of life history strategies in aquatic ecosystems: Ecosim II. *Ecosystems*, **3**: 70–83, 2000.

Warren PH. Making connections in food webs. *Trends in Ecology and Evolution*, **4**: 136–140, 1994.

Wasserman S, Faust K. *Social Network Analysis: Methods and Applications*. Cambridge University Press, Cambridge, UK, 1994.

Watts D, Strogatz S. Collective dynamics of small world networks. *Nature*, **393**: 440–442, 1998.

Wei W. *Biodiversity Analysis on Arthropod and Weed Communities in Paddy Rice Fields of Pearl River Delta*. Master Degree Dissertation. Sun Yat-sen University, China, 2010.

Wiegand T, Moloney KA, Naves J, Knauer F. Finding the missing link between landscape structure and population dynamics: a spatially explicit perspective. *American Naturalist*, **154**: 605–627, 1999.

Williams RJ. Simple MaxEnt models explain food web degree distributions. *Theoretical Ecology*, **3**: 45–52, 2010.

Wilson MA, Bhalla US, Uhley JD, *et al.* Genesis: a system for simulating neural networks. NIPS, 485–492, 1988.

Wilson EO. *The Diversity of Life*. Harvard University Press, Cambridge, MA, USA, 1992.

Winemiller KO, Pianka ER, Vitt LJ, *et al.* Food web laws or niche theory? Six independent empirical tests. *American Naturalist*, **158**: 193–199, 2001.

Wolfram S. 2002. *A New Kind of Bcience*. Wolfram Media, USA.

Xie NG, Peng FR, Ye Y, *et al.* Research on evolution of cooperation among biological system based on Parrondo's Paradox Game. *Journal of Anhui University of Technology*, **27**(2): 167–174, 2010.

Yan PF, Zhang CS. *Artificial Neural Networks and Computation of Simulated Evolution*. Tsinghua University Press, Beijing, China, 2000.

Yodzis P. The connectance of real ecosystems. *Nature*, **284**: 544–545, 1980.

Zhang WJ. Computer inference of network of ecological interactions from sampling data. *Environmental Monitoring and Assessment*, **124**: 253–261, 2007a.

Zhang WJ. *Methodology on Ecology Research*. Sun Yat-sen University Press, Guangzhou, China, 2007b.

Zhang WJ. *Computational Ecology: Artificial Neural Networks and Their Applications*. World Scientific, Singapore, 2010.

Zhang WJ. Supervised neural network recognition of habitat zones of rice invertebrates. *Stochastic Environmental Research and Risk Assessment*, **21**: 729–735, 2007c.

Zhang WJ. Pattern classification and recognition of invertebrate functional groups using self-organizing neural networks. *Environmental Monitoring and Assessment*, **130**: 415–422, 2007d.

Zhang WJ. Constructing ecological interaction networks by correlation analysis: hints from community sampling. *Network Biology*, **1**(2): 81–98, 2011a.

Zhang WJ. Network Biology: an exciting frontier science. *Network Biology*, **1**(1): 79–80, 2011b.

Zhang WJ. A Java algorithm for non-parametric statistic comparison of network structure. *Network Biology*, 1(2): 130–133, 2011c.

Zhang WJ. A Java program to test homogeneity of samples and examine sampling completeness. *Network Biology*, 1(2): 127–129, 2011d.

Zhang WJ, Barrion AT. Function approximation and documentation of sampling data using artificial neural networks. *Environmental Monitoring and Assessment*, **122**: 185–201, 2006.

Zhang YT, Fang KT. *Introduction to Multivariable Statistics*. Science Press, Beijing, China, 1982.

Zhang WJ, Gu DX. A non-linear partial differential equation to describe spatial and temporal changes of insect population. *Ecologic Science*, **20**(4): 1–7, 2001.

Zhang WJ, Schoenly KG. IRRI Biodiversity Software Series. II. COLLECT1 and COLLECT2: Programs for Calculating Statistics of Collectors' Curves. *IRRI Technical Bulletin No.2*. International Rice Research Institute, Manila, Philippines, 1999.

Zhang WJ, Schoenly KG. IRRI Biodiversity Software Series. III BOUNDARY: a program for detecting boundaries in ecological landscapes. *IRRI Technical Bulletin No.3*. International Rice Research Institute, Manila, Philippines, 1999.

Zhang WJ, Wei W. Spatial succession modeling of biological communities: a multi-model approach. *Environmental Monitoring and Assessment* **158**: 213–230, 2009.

Zhang WJ, Zhan CY. An algorithm for calculation of degree distribution and detection of network type: with application in food webs. *Network Biology*, 1(3–4): 159–170, 2011.

Zhang WJ, Zhang RJ, Gu DX. An algorithm and network implementation of clustering analysis with randomized statistical testing. *Computer Engineering and Science*, **28**(12): 74–76, 2006.

Zhang WJ, Zhang XY. Neural network modeling of survival dynamics of holometabolous insects: a case study. *Ecological Modelling*, **211**: 433–443, 2008.

Zhou WG. A feld survey on paddy rice arthropod biodiversity in northern guangzhou. Master Degree Dissertation. Sun Yat-sen University, China, 2007.

Zimmer Carl. 100 trillion connections. *Scientific American*, **304**(1): 45–49, 2011.

Zorach AC, Ulanowicz RE. Quantifying the complexity of low networks: how many roles are there? *Complexity*, **8**: 68–76, 2003.

Index